Forensic Analysis of Biological Evidence

A Laboratory Guide for Serological and DNA Typing

J. Thomas McClintock, PhD
DNA Diagnostics, Inc
Lynchburg, Virginia

CRC Press
Taylor & Francis Group
Boca Raton London New York

CRC Press is an imprint of the
Taylor & Francis Group, an **informa** business

CRC Press
Taylor & Francis Group
6000 Broken Sound Parkway NW, Suite 300
Boca Raton, FL 33487-2742

Printed on acid-free paper
Version Date: 20131101

International Standard Book Number-13: 978-1-4665-0456-1 (Paperback)

Library of Congress Cataloging-in-Publication Data

McClintock, J. Thomas, author.
Forensic analysis of biological evidence : A laboratory guide for serological and DNA typing / J. Thomas McClintock.
pages cm
Summary: "Whether a stand-alone laboratory manual or in conjunction with Richard Li's Forensic Biology, Second Edition, this book provides more than a dozen exercises covering analysis of biological evidence. Originally published as Forensic DNA Analysis: A Laboratory Manual, this greatly revised work offers updated exercises and protocols for all kinds of DNA and serological analyses with delineated objectives, step-by-step procedures, and laboratory supplies including advanced instrumentation. It provides a glossary of terms, 14 general information tables, and ancillary instructor's materials along with background material on DNA, downloadable exercises forms, and video instruction for self-teaching"-- Provided by publisher.
ISBN 978-1-4665-0456-1 (pbk.)
1. DNA--Analysis--Laboratory manuals. 2. DNA fingerprinting--Laboratory manuals. 3. Forensic genetics--Laboratory manuals. I. Title.

RA1057.55.M33 2014
614'.12--dc23 2013033005

Visit the Taylor & Francis Web site at
http://www.taylorandfrancis.com

and the CRC Press Web site at
http://www.crcpress.com

Contents

List of Figures

List of Tables

Preface

DNA typing has revolutionized criminal investigations and has become a powerful tool in the identification of individuals in criminal and paternity cases. In the past few years, the general public has become familiar with forensic DNA typing based on exposure from media coverage (e.g., the O.J. Simpson trial, the President Clinton and Monica Lewinsky scandal, and identification of individuals killed in the September 11, 2001 attack on the World Trade Center (WTC) in New York City and the Pentagon in Arlington, Virginia) and television (e.g., *Forensic Files*, *CSI: Miami*). Although these cases have generated widespread media attention, they represent only a small fraction of the thousands of forensic DNA and paternity cases that are conducted by public and private laboratories in the United States and abroad.

The purpose of this laboratory manual is to introduce the student to the science of serological analysis and DNA typing methods by focusing on basic techniques used in forensic DNA laboratories. This laboratory manual is designed to provide the student with a fundamental understanding of serological and forensic DNA analysis as well as a thorough background of the molecular techniques used to determine an individual's identity or parental lineage. This manual is intended to challenge the student with the methodology of the investigation in DNA typing, develop an understanding of the scientific principles involved in serology and DNA analysis, and be able to analyze and interpret the data that are generated in each exercise with clarity and confidence.

The exercises in this laboratory manual have been organized to first provide an overview of forensic DNA analysis, the sources or types of biological material used in DNA analysis, and then the background principles and practical methodology for a specific serological analysis and DNA typing technique. In some exercises, the protocols have been adapted from methods and protocols used in federal, state, and private forensic laboratories. Each exercise is designed to simulate human forensic testing but can also be used to simulate a wide range of applications for genetic analysis. The actual scenario employed in each exercise is up to the discretion of the course instructor. Lastly, an extensive glossary has been included to assist students with terminology used in the forensic analysis of biological evidence, as well as basic terms used in molecular biology.

Compiled below is a brief history of forensic DNA typing. Since DNA testing was first introduced in the United States in 1986, it has been used in thousands of cases. However, the list below highlights specific events or developments in forensic DNA analysis as well as those cases brought to the attention of the general public by media exposure.

Brief History of Forensic DNA Typing

- 1980: Ray White describes first polymorphic restriction fragment length polymorphism (RFLP) markers.
- 1985: Alec Jeffreys discovers multilocus variable number tandem repeat (VNTR) probes.
- 1985: First paper on polymerase chain reaction (PCR).

- 1986: DNA testing goes public (Cellmark and Lifecodes).
- 1986: First RFLP case in the United States (*Florida v. Tommy Lee Andrews*).
- 1988: FBI starts DNA casework (RFLP).
- 1989: The Technical Working Group on DNA Analysis Methods (TWGDAM) was established.
- 1991: First short tandem repeat (STR) paper.
- 1992: National Research Council I report *DNA Technology in Forensic Science*.
- 1993: First STR kit available.
- 1995: Forensic Science Service (FSS) starts UK DNA database.
- 1995: O.J. Simpson trial; public becomes aware of DNA.
- 1996: National Research Council (NRC) II report *The Evaluation of Forensic DNA Evidence*.
- 1996: First use of mitochondrial DNA test in a U.S. criminal trial (*Tennessee v. Ware*).
- 1998: FBI launches Combined DNA Index System (CODIS) database.
- 1998: Establishment of quality assurance standards for forensic DNA testing laboratories through the DNA Advisory Board.
- 1998: Kenneth Starr investigates allegations of President Clinton's sexual relationship with a White House intern, Monica Lewinsky.
- 1999: Multiplex STRs are validated.
- 1999: The decision in *State v. Ware* (1996) was upheld by an appellate court.
- 2002: Division of Forensic Science Laboratory in the Commonwealth of Virginia became the first state laboratory to mark 1,000 "cold hits" from its DNA database.
- 2003: A field DNA test was completed to provide preliminary confirmation of the identification of Saddam Hussein less than 24 hours after his capture. A full test performed in the laboratory provided confirmation.
- 2003: The National Institute of Standards and Technology develops a "mini-STR assay" to allow the remains from 16 additional victims from the September 11, 2001 attacks on the World Trade Center to be positively identified.
- 2006: Members of the Duke University men's lacrosse team are arrested and accused of raping a female exotic dancer. Samples are collected from the dancer and the men's lacrosse team for DNA analysis.
- 2006: DNA testing failed to connect any members of the Duke University men's lacrosse team to the alleged sexual assault of an exotic dancer.
- 2007: Prosecutor handling the Duke case is forced to recuse himself. North Carolina's attorney general declared three former Duke University lacrosse players who had been accused of gang raping a stripper innocent of all charges, ending a prosecution that provoked bitter debate over race, class, and the tactics of the Durham County district attorney.
- 2007: Applied Biosystems introduced the first DNA testing kit for analyzing degraded or limited DNA.
- 2007: FSS in the UK uses laser microdissection (LMD), which enables single cells to be extracted from a microscope slide, with fluorescence *in situ* hybridization (FISH), a method to highlight chromosomes, to distinguish between male (XY chromosomes) and female (XX chromosomes) cells.
- 2008: Idaho National Laboratory develops a method for identifying people through antibody analysis, a technique reported to be inexpensive, faster, and simpler than DNA analysis.
- 2009: Standards for forensic DNA testing laboratories and standards for DNA databasing laboratories were updated by the DNA Advisory Board and implemented by the FBI.
- 2010: Applied Biosystems' AmpFlSTR MiniFiler, a forensic DNA kit designed to obtain DNA results from compromised or degraded samples, receives approval for inclusion in the National DNA Index System (NDIS).
- 2010: Two new Applied Biosystems forensic DNA test kits (Identifiler Direct and Identifiler Plus) receive approval for use by the National DNA Index System (NDIS).

- 2011: The Commonwealth of Virginia becomes the third state to approve familial DNA searches in the DNA database for individuals who might be closely related to people whose DNA is found at a crime scene.
- 2011: Osama bin Laden killed by U.S. forces. DNA tests confirmed that the body recovered from the 1-acre compound in Abbottabad, Pakistan was bin Laden.
- 2012: A forensic scientist linked DNA on medical waste to Roger Clemens in the perjury trial of major league baseball's seven-time Cy Young Award winner. The "Rocket" was ultimately acquitted on all charges by a jury that decided he did not lie to Congress when he denied using performing-enhancing drugs.
- 2012: Scientists develop a forensic test that can predict both the hair and eye color of a possible suspect using DNA left at a crime scene.
- 2013: Life Technologies introduces a portable DNA testing system that can identify whether a specimen contains human DNA that can be analyzed, and the person's gender, in 75 minutes.

It is hoped that this laboratory manual will develop the student's curiosity and confidence to further explore questions and issues involving forensic science investigations. I look forward to teaching you the techniques and applications in forensic analysis of biological evidence.

J. Thomas McClintock, PhD

DNA Diagnostics, Inc.
Lynchburg, VA
www.DNADiagnosticsInc.com

About the Author

DNA Diagnostics, Inc. was founded in 1993 by **Dr. J. Thomas McClintock**, a forensic scientist who specializes in the analysis of DNA and the use of insects in legal investigations. Dr. McClintock has evaluated and provided expert guidance on close to 300 criminal and paternity cases, in both the public and the private sector, where DNA analysis was performed for individual identification or parentage verification. Dr. McClintock has taught numerous training seminars that focus on the principles and methodologies used in forensic DNA typing as well as workshops designed to train prosecutors, investigators, and law enforcement officials in (1) crime scene preservation, (2) evidence collection, (3) chain of custody issues, (4) DNA evidence handling and analysis, (5) specialized trial tactics, and (6) effective presentation of DNA evidence to judges and juries.

Dr. McClintock is a professor in the Department of Biology and Chemistry at Liberty University in Lynchburg, Virginia, where he teaches undergraduate courses in microbiology. He previously held a faculty position in the Department of Molecular and Microbiology at George Mason University (GMU) in Fairfax, Virginia, where he taught graduate courses in forensic DNA analysis, forensic entomology, and forensic sciences. The course in forensic DNA analysis focused on current laboratory methods and applications in forensic DNA profiling and effective presentation of DNA evidence at trial. His previously published laboratory manual, entitled *Forensic DNA Analysis: A Laboratory Manual* (2008, CRC Press/Taylor & Francis), presents an overview of the current techniques and methods commonly used in forensic DNA typing, as well as the interpretation of DNA evidence, which is crucial for today's criminalist.

In 2008, Dr. McClintock was invited to appear on the *Nancy Grace* show (www.CNN.com/CNN/Programs/Nancy.Grace) to provide his insights on the use of DNA testing in the investigation of the missing 3-year-old Florida girl Caylee Anthony. The complete transcripts for both shows are available at http://www.Edition.cnn.com/TRANSCRIPTS/0808/22/ng.01.html and http://www.Transcripts.cnn.com/TRANSCRIPTS/0809/26/ng.01.html.

Dr. McClintock received a bachelor of science degree from the Department of Biology at James Madison University, Harrisonburg, Virginia, a master's of science degree from the Department of Entomology, and a doctorate of philosophy degree from the Department of Microbiology at the University of Maryland, College Park. After completing a postdoctoral fellowship, Dr. McClintock became group leader/senior staff scientist at Digene Diagnostics, Inc., where he directed the research on the development and use of DNA probes for the detection and diagnosis of human pathogens.

For more information, see http://www.DNADiagnosticsInc.com or contact Dr. McClintock at DNADiagnosticsInc@gmail.com.

List of Acronyms/Abbreviations

ABO	Blood group system
AK	Adenylate kinase
ALS	Alternate light source
AP	Acid phosphatase
CE	Capillary electrophoresis
$CHCl_3$	Chloroform
CODIS	Combined DNA Index System
CSF1PO	Locus or site on chromosome 5
D1S80	Locus or site on chromosome 1
D7S80	Locus or site on chromosome 7
DNA	Deoxyribonucleic acid
DTT	Dithiothreitol
EAP	Erythrocyte acid phosphatase
EDTA	Ethylenediamine tetraacetic acid
EtOH	Ethanol
FBI	Federal Bureau of Investigation
FISH	Fluorescence *in situ* hybridization
FSS	Forensic Science Service
FTA	Collection card for biological samples
GC	Group-specific component; locus on chromosome 4
GYPA	Glycophorin A; locus on chromosome 4
HaeIII	Restriction endonuclease
HBGG	Hemoglobin G gammaglobin
HCL	Hydrochloric acid
HeLa	Human cell line
HepG	Human cell line
HindIII	Restriction endonuclease
HLA DQA1	Human leukocyte antigen; locus on chromosome 6
HVI	Hypervariable region 1 in mtDNA
HVII	Hypervariable region 2 in mtDNA

INC	Inconclusive
K562	Human cell line
KS	Kernechtrot staining solution
LCG	Leucomalachite green
LDLR	Low-density lipoprotein receptor; locus on chromosome 19
LMD	Laser microdissection
mtDNA	Mitochondrial DNA
NaCl	Sodium chloride
NDIS	National DNA Index System
ng	Nanogram; one billionth of a gram
NRC	National Research Council
PBS	Phosphate buffer saline
PCR	Polymerase chain reaction
PGM	Phosphoglucomutase
PIC	Picroindigocarmine
PM	Polymarker
RBC	Red blood cell
REN	Restriction endonuclease
RFLP	Restriction fragment length polymorphism
RNA	Ribonucleic acid
RSID	Rapid stain identification test
SDS	Sodium dodecyl sulfate
SNP	Single-nucleotide polymorphism
STR	Short tandem repeat
SWGDAM	Scientific Working Group on DNA Analysis Methods
TAE	Tris-acetate buffer
TE	Tris-EDTA buffer
THO1	Locus or site on chromosome 11
TNE	Tris–sodium chloride–EDTA buffer
TPOX	Locus or site on chromosome 2
TWGDAM	Technical Working Group of DNA Analysis Methods
VNTR	Variable number tandem repeat
WBC	White blood cell
Y-STR	Y chromosome short tandem repeat

Safety and Other Considerations in the Laboratory

1. There will be no eating, drinking, smoking, applying cosmetics, or handling contacts in the laboratory at *any* time.
2. There will be no pipetting by mouth. Use a pipettor at all times.
3. Minimize splashing and production of aerosols.
4. Store all books, backpacks, cell phones and other electronic devices, purses, and coats in the cabinet of your laboratory bench (or designated area). Only your laboratory notebook should be on the bench.
5. Do not place pencils, pens, or any other object into your mouth while in the laboratory.
6. *Never* take any reagents, samples, or cell cultures out of the laboratory.
7. Notify the laboratory instructor immediately of any spills, of any accidents, or if you cut or injure yourself.
8. In most instances you will be wearing disposable gloves. Be extra careful when handling reagents or chemicals to eliminate skin contact. If working with samples thought to contain biological material (i.e., blood) wear goggles with a particle mask as well as gloves.
9. Wash your hands at the beginning and at the end of the laboratory exercises.
10. Clean your laboratory bench with dilute alcohol (70%) or with a 10% bleach solution before you begin work and when your have completed the laboratory exercise.
11. Laboratory coats are not required. However, in forensic laboratories, laboratory coats must be worn as well as disposable gloves.
12. Children should not be allowed in the laboratory.
13. Familiarize yourself with the location of the eye wash station, the fire extinguisher, and the fire blanket.
14. Dispose of all materials potentially contaminated with biological material into a disposable biohazard container.

Chapter 1

An Overview of Forensic DNA Analysis

Everyone has a unique set of fingerprints. As with a person's fingerprint, no two individuals share the same genetic makeup. This genetic makeup, which is the hereditary blueprint imparted to us by our parents, is stored in the chemical deoxyribonucleic acid (DNA), the basic molecule of life. Examination of DNA from individuals, other than identical twins, has shown that variations exist and that a specific DNA pattern or profile can be associated with an individual. These DNA profiles have revolutionized criminal investigations and have become powerful tools in the identification of individuals in criminal and paternity cases.

Restriction Fragment Length Polymorphism

The first widespread use of DNA tests involved restriction fragment length polymorphism (RFLP) analysis, a test designed to detect variations in human DNA. In the RFLP method, DNA is isolated from a biological specimen (e.g., blood, semen, vaginal swabs) and cut by a restriction endonuclease (e.g., an enzyme such as HaeIII) into pieces called restriction fragments. The DNA fragments are separated by size into discrete bands by gel electrophoresis, transferred onto a membrane by Southern (1975) blotting, and identified using probes (known DNA sequences that are "tagged" with a chemical tracer). The resulting DNA profile is visualized by exposing the membrane to x-ray film, allowing the scientist to determine which specific fragments the probe identified among the thousands in a sample of human DNA (Figure 1.1). A match is made when similar DNA profiles are observed between an evidentiary sample and those from a known sample (e.g., DNA from a victim or suspect). A determination is then made as to the probability that a person selected at random from a given population would match the evidence sample as well as the suspect. The entire analysis may require several weeks for completion.

Polymerase Chain Reaction-Based Tests

In instances when the evidentiary sample contains an insufficient quantity of DNA or the DNA is degraded, a polymerase chain reaction (PCR)-based test is used to obtain a DNA profile. The PCR-based tests generally provide rapid results that can serve as an alternative or as a complement to other DNA tests. The first step in the PCR process involves the isolation of DNA from a biological specimen (e.g., blood, semen, saliva, fingernail clippings). Next, the PCR amplification technique is used to produce millions of copies of a specific portion of a targeted DNA segment. The PCR amplification procedure is comparable to a photocopying machine only at the molecular level. The amplified PCR products are then either identified by the addition of known DNA probes (e.g., human leukocyte antigen (HLA) DQA1 and polymarker (PM) test kits; Figures 1.2 and 1.3, respectively) or separated by gel electrophoresis (D1S80, short tandem repeat (STR), and amelogenin (gender) analyses) followed by chemical staining. Such detection procedures eliminate the need for critically sensitive DNA probes, thus reducing the analysis time from several weeks to 24 to 48 h.

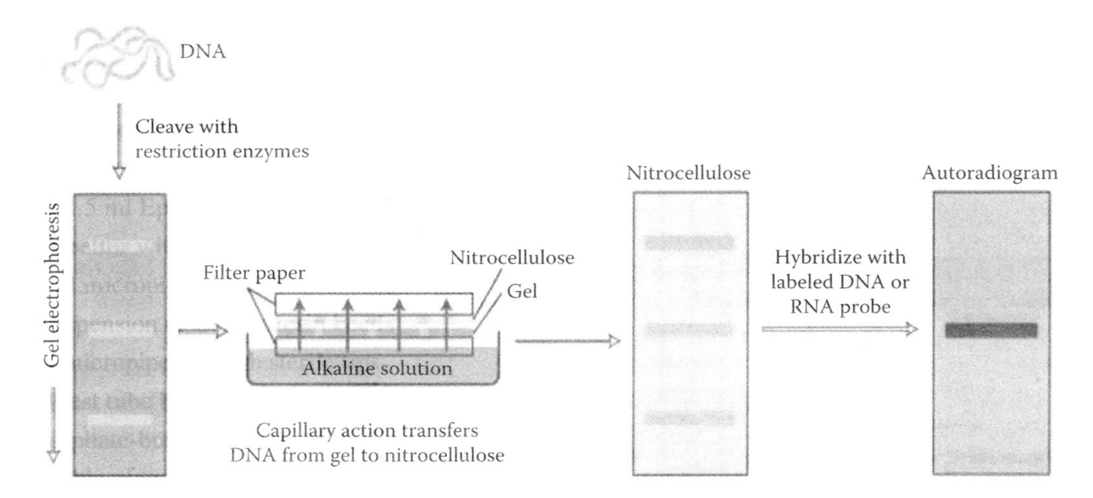

Figure 1.1
An overview of the RFLP procedure. (From the University of Strathclyde in Glasgow's Internet home page; http://www.strathclyde.ac.uk/~dfs99109/BB211/RecombDNAtechlect2.html. With permission.))

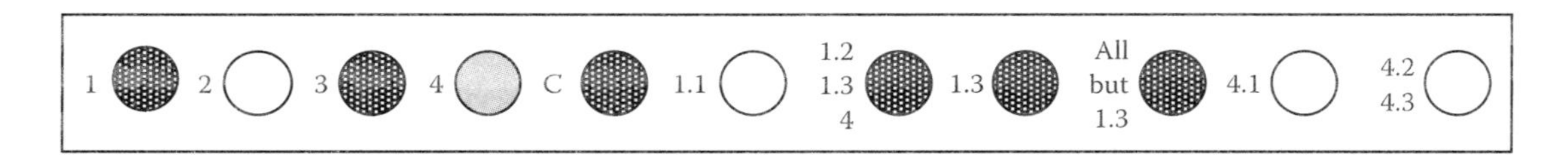

Figure 1.2
Human leukocyte antigen (HLA) DQA1 typing results. Prior to hybridization the nylon typing strips are colorless. Following hybridization, a colorless substrate is added, and if the amplified DNA binds to the probe, the enzyme will convert the colorless substrate to a blue precipitate. The pattern of the "blue dots" is an example of a DQA1 typing strip that was hybridized with amplified DNA from a person with a DQA1 type of 1.3, 3. The typing results are interpreted from left to right. In this example, a faint response is observed at the C dot or control dot. There is a response at the 1 allele, a response at the 3 allele (confirming a 3 allele), a response at the 1.2, 1.3, 4 alleles, a response at the 1.3 allele (confirming a 1.3 allele), and a response at the "all but 1.3" alleles. Based on these responses, the DQA1 type or profile is 1.3, 3. The circles surrounding an area on the nylon strip are only present in these illustrations to indicate the location of the DNA probes.

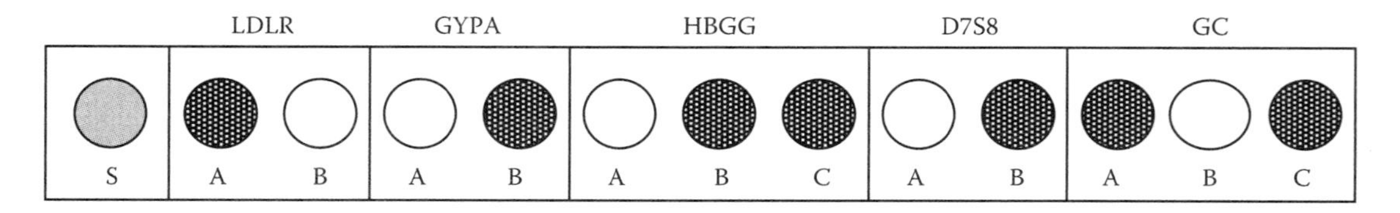

Figure 1.3
Polymarker (PM) typing results. Prior to hybridization the nylon typing strips are colorless. The DNA probes are immobilized onto the nylon strip and identified with the corresponding locus (e.g., LDLR, GYPA, etc.). The circles surrounding an area on the nylon strip are only present in these illustrations to indicate the location of the DNA probes. Each locus will have two or three invisible dots that represent the allelic variations at that locus. In this example, a faint response is observed at the S dot or control dot. There is a response at the A allele of the LDLR locus, a response at the B allele of the GYPA locus, two responses at the B and C alleles of the HBGG locus, a response at the B allele of the D7S8 locus, and a response at the A and C alleles of the GC locus. Based on these responses, the AmpliType PM type or profile is AA/BB/BC/BB/AC or A/B/BC/B/AC.

The resulting DNA profiles are routinely interpreted by direct comparison to DNA standards. Probability calculations are determined based upon classical population genetic principles.

Mitochondrial DNA Analysis

Mitochondrial DNA (mtDNA) typing is increasingly used in human identity testing when biological evidence may be degraded, when quantities of the samples in question are limited, or when nuclear DNA typing is not an option. Biological sources of mtDNA include hairs, bones, and teeth. In humans, mtDNA is inherited strictly from the mother. Consequently, mtDNA analysis cannot discriminate between maternally related individuals (e.g., mother and daughter, brother and sister). However, this unique characteristic of mtDNA is beneficial for missing person cases when mtDNA samples can be compared to samples provided by the maternal relative of the missing person.

In humans, the mtDNA genome is approximately 16,569 bases (A, T, G, and C) in length, containing a "control region" with two highly polymorphic regions. These two regions, termed hypervariable region 1 (HV1) and hypervariable region 2 (HV2), are 342 and 268 base pairs (bp) in length, respectively, and are highly variable within the human population. This sequence (the specific order of bases along a DNA strand) variability in either region provides an attractive target for forensic identification studies. Moreover, since human cells contain several hundred copies of mtDNA, substantially more template DNA is available for amplification than nuclear DNA.

Mitochondrial DNA typing begins with the extraction of mtDNA followed by PCR amplification of the hypervariable regions. The amplified mtDNA is purified, subjected to sequencing (Sanger et al., 1977), with the final products containing a fluorescently labeled base at the end position. The products from the sequencing reaction are separated, based on their length, by gel or capillary electrophoresis. The resulting sequences or profiles are then compared to sequences of a known reference sample to determine differences and similarities between samples (Anderson et al., 1981; Andrews et al., 1999). Samples are not excluded as originating from the same source if each base (A, T, G, or C) at every position along the hypervariable regions is similar. However, due to the size of the mtDNA database and to the unknown number of mtDNA sequences in the human population, a reliable frequency estimate is not provided. Consequently, mtDNA sequencing is becoming known as an exclusionary tool as well as a technique to complement other human identification procedures.

Chapter 2

Types of Biological Samples

With the exception of red blood cells, DNA is found in every human cell. Consequently, DNA is present in a variety of body fluids and tissues that have been demonstrated to be suitable for DNA typing. However, if the sample or evidence collection is performed improperly, the sample's integrity may be compromised, leading to contamination or degradation. Improper handling procedures during storage and transport from the crime scene to the laboratory can result in samples unfit for analysis. The importance of sample integrity cannot be overemphasized since ambiguous data/information can compromise the investigation or the outcome of the case.

In the past, DNA typing tests such as restriction fragment length polymorphism (RFLP) were successful in generating complete DNA profiles provided that adequate and nondegraded samples were utilized. The introduction of the polymerase chain reaction (PCR) in the mid- to late 1980s extended the range of possible samples available for DNA analysis regardless of their condition. Some of the biological samples that have been tested successfully with PCR-based typing methods are listed in Table 2.1. The minimum or corresponding amount of DNA available from each biological sample is also shown.

Prior to DNA isolation and typing, biological samples must first be collected either from a known contributor (victim or suspect) or from the crime scene (evidentiary sample). Once the sample is collected, the DNA is extracted and subjected to PCR analysis. In general, the PCR procedure typically requires as little as 1 ng (1 billionth of a gram) of high molecular weight genomic DNA. DNA thought to be degraded can also be subjected to PCR analysis since intact, high molecular weight DNA is not necessary to generate a complete DNA profile. Table 2.2 illustrates the types of physical evidence collected at crime scenes, the various locations of the DNA, and the biological source.

TABLE 2.1
DNA Contents of Biological Samples

Biological Sample/Source	Approximate DNA Content
Liquid blood	20–40 μg/ml (1 μl from 4 to 11 × 10^3 WBC)
Bloodstain (1 cm^2)	250–500 ng
Semen	150,000–300,000 ng/ml
Postcoital vaginal swab	0–3,000 ng/ml
Saliva	1,000–10,000 ng/ml
Oral swab	1,000–1,500 ng
Hair roots	1–750 ng/plucked hair root
Hair (shed)	1–12 ng/hair
Urine	1–20 ng/ml
Bone	3–10 ng/mg bone
Tissue (15 mg)	3–15 μg/mg
Fibroblast cell line	6.5 μg/1 × 10^6 cells

TABLE 2.2
Physical Evidence Collected at Crime Scenes

Physical Evidence	Location of DNA	Biological Source of DNA
Used cigarette	Cigarette butt	Saliva
Toothpick	Tips	Saliva
Stamps and envelopes	Licked area	Saliva
Bottle or can	Mouthpiece	Saliva
Used condom	Inside/outside surface	Semen, vaginal or rectal cells
Blanket, sheet, or pillow	Surface	Semen, sweat, hair, saliva, urine
Bite mark	Clothing or person's skin	Saliva
Fingernail	Scrapings	Blood, skin, sweat
Tape or ligature	Inside/outside surface	Skin/surface
Bullet	Outside surface	Blood, tissue, skin
Clothing	Surface area	Blood, sweat, semen
Hat, mask, bandanna	Inside surface	Sweat, hair, dandruff
Knife, bat, or similar object	Outside surface/handle	Blood, skin, tissue, sweat

Biological evidence will attain its full forensic value only when the DNA types can be compared to known profiles obtained from the victims and suspects. The following is a brief description of the various biological samples suitable for DNA typing with accompanying guidance for collection and storage.

Whole Blood

Whole blood from a known source should be collected in a sterile tube containing the preservative ethylenediamine tetraacetic acid (EDTA). In addition to acting as a preservative, EDTA also inhibits the activity of enzymes that are responsible for degrading DNA. Tubes containing blood samples should be stored at refrigerated (for short periods of time) or frozen (for long-term storage) temperatures. Whole blood can also be "spotted" onto an FTA collection card (an absorbent cellulose-based paper that contains chemical substances to inhibit bacterial growth and to protect the DNA from enzymatic degradation), allowed to dry, and stored at room temperature for several years (Figure 2.1).

Bloodstains or Mixed Stains

Garments or clothing containing stains are packaged in a suitable manner (e.g., bag or box) and transported to the laboratory and stored until analysis. It should be noted that all stained material should be dried thoroughly prior to packaging and submission to the laboratory for analysis. Short-term storage should be at room temperature in a humidity-controlled room out of direct sunlight. For long-term storage, samples should be stored in a low-temperature frost-free freezer. Once the packaged material arrives at the laboratory, the removal of the stain from the item is performed by the forensic DNA analyst. Stains comprising a size of approximately 5 mm^2 (or about the size of a quarter) or greater, and with a volume of less than 5 μl, have been successfully analyzed by current DNA typing techniques.

Hairs

In general, forensic hair analysis involves either head or pubic hair (Figure 2.2). The collection of 12–24 full-length hairs from the scalp will provide more than enough material for analysis. Hair samples that contain an

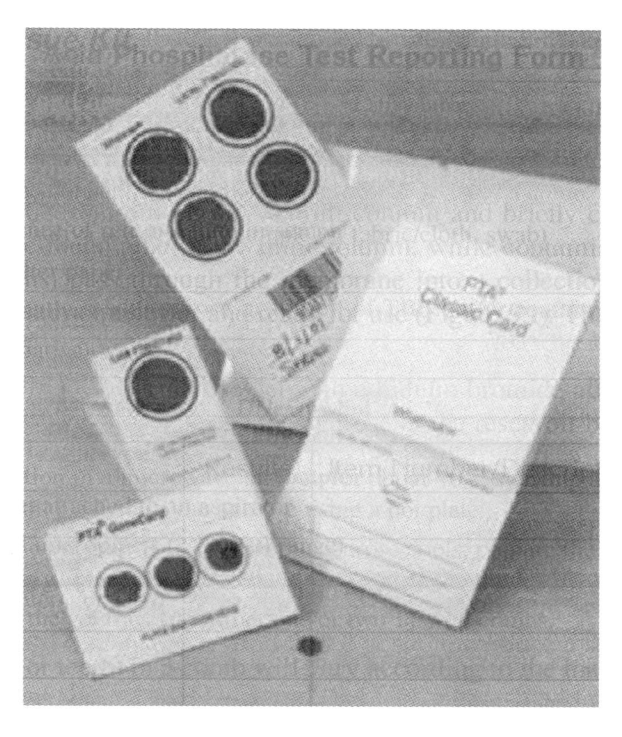

Figure 2.1
FTA mini-cards for collection and storage. Whatman FTA cards are made of a chemically treated paper that extracts the DNA from a blood sample. A drop of blood is spotted on the card and the DNA is instantly captured. FTA cards can be mailed without special handling to a laboratory for analysis, or stored indefinitely at room temperature. (Courtesy of Whatman Ltd.)

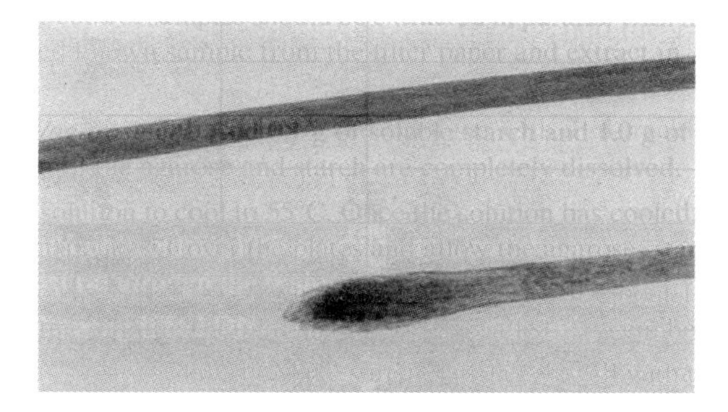

Figure 2.2
Photomicrograph of human head hair and root (bottom). (From FBI website: www.fbi.gov.)

intact root will provide enough nuclear DNA for short tandem repeat (STR) typing (see Exercise 10). A hair shaft contains sufficient mitochondrial DNA (mtDNA) for successful mtDNA typing (see Exercise 13). As with stains, short-term storage may be at room temperature in a humidity-controlled room and out of direct sunlight. For long-term storage of hair samples, a low-temperature non-frost-free freezer is recommended.

Swabs from Biological Material or Inanimate Objects

DNA has been successfully analyzed from swabs containing biological material (e.g., buccal cells from the inside of the cheek or epithelial cells from a vaginal swab) to swabs of various inanimate objects such as cigarettes, envelopes, soda cans, and stamps. Using a sterile and moistened cotton swab, the area in question is swabbed, the swab dried, and then placed in a container for storage or placed in a vial containing a small volume of sterile solution such as 1× TE buffer for short-term storage. Swabs containing the biological material for analysis are stored as recommended, as with stains.

Bone, Teeth, and Tissue

In some instances, the probability of obtaining sufficient quantity or quality of biological material such as blood or semen for DNA typing will be low. This insufficient quality/quantity of material may be due to sample degradation, availability, or even accessibility. In these cases, samples such as bones, teeth, skin, and muscle tissue will usually provide sufficient DNA for analysis (Figure 2.3). Generally, a 1 cm^2 section (slightly larger than a dime) of such biological material is suitable for testing. Following collection, the sample should be frozen and transported to the laboratory on ice. Upon arrival at the laboratory, the samples should be kept frozen until the DNA typing analysis begins. To ensure sample integrity, avoid multiple freeze-thaw conditions.

Paraffin-Embedded Tissues, Smears, Slides

When known biological samples are unavailable, but needed, investigation into local medical facilities may yield a source: specimens collected from biopsies or surgical procedures that were processed, analyzed, and stored. Such specimens may include vaginal/pap smears or histological sections. Histological sections contain samples fixed in formalin, processed, and embedded in paraffin (Figure 2.4). The paraffin "blocks" are sectioned using a microtome and placed on polylysine-coated slides. Usually, a minimum of one block, smear, or slide is typically necessary for a successful DNA typing analysis. Histological sections, smears, or slides may be stored at room temperature indefinitely prior to analysis.

Isolation of DNA from paraffin-embedded tissues is an arduous task for DNA analysts. High yields and purity are the main problems using DNA extracted from such samples. However, new methods have been developed that avoid the use of xylene for paraffin removal, which eliminates extra handling steps that can lead to sample loss. Consequently, maximum DNA yields from as little as a single histological section of tissue suitable for DNA typing analysis are now possible (Kennedy, 2009).

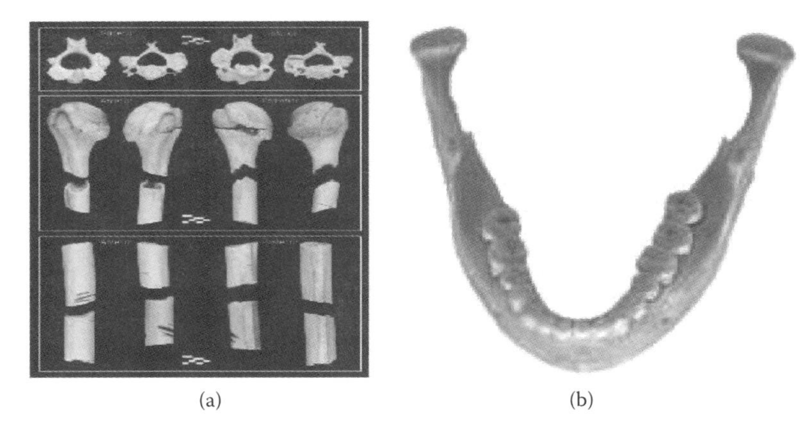

(a) (b)

Figure 2.3
Photographs of (a) human bones and (b) a lower jawbone with teeth prepared for DNA analysis. (Courtesy of Steven A. Symes, Ph.D., Mercyhurst College.)

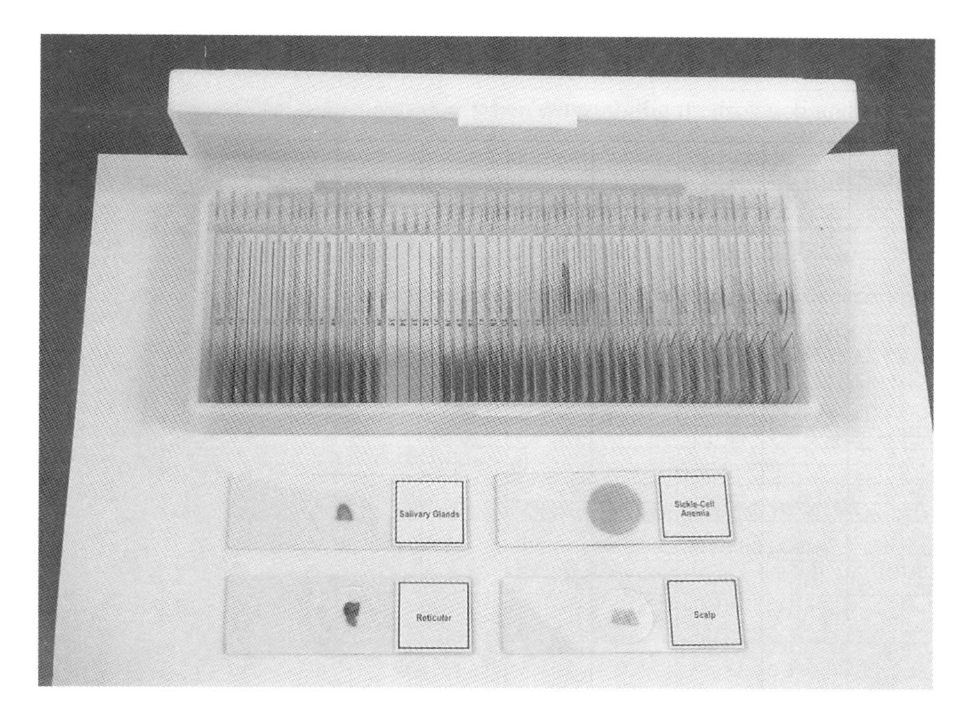

Figure 2.4
Histological samples on slides. Slides containing paraffin-embedded tissue sections or vaginal or pap smears can be valuable sources of biological material when known samples are unavailable. (Courtesy of DR Instruments, Inc., www.drinstruments.com.)

Semen/Sperm

Sperm specimens, collected from a vaginal swab, will contain epithelial (skin) cells from the female and, in some instances, from the male (Figure 2.5). Sperm can be preferentially separated from the rest of the material in such a mixture using specific extraction methods discussed in Exercise 6 (see "Differential Extraction" section). Samples or stains thought to or have been shown to contain only spermatozoa can be collected and processed as described above (see "Bloodstains or Mixed Stains" section). Sperm specimens should be stored frozen prior to analysis. To ensure sample integrity avoid multiple freeze-thaw conditions.

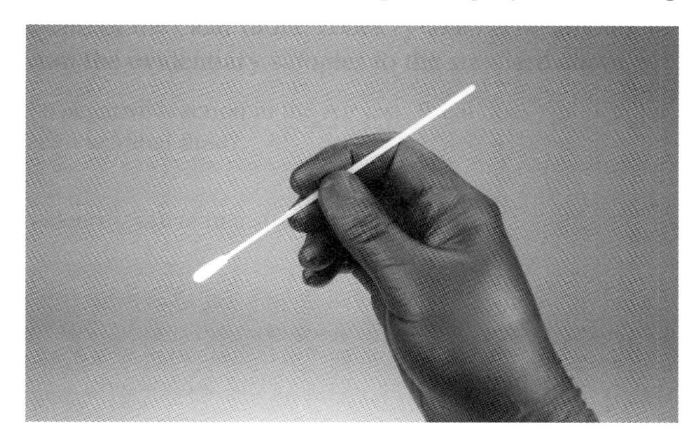

Figure 2.5
Flocked swabs for the collection of semen. (From Michael J. Grimm, EVIDENT, Inc., http://www.evidentcrimescene.com/data/dna/dna.html. With permission.)

Urine

Similar to perspiration and sebaceous oils, urine, when concentrated, will contain a sufficient amount of epithelial cells to generate a DNA profile. Ideally, a minimum sample volume of 10 ml is required for analysis, with an optimum approaching 30 ml (Figure 2.6). However, a sufficient number of cells might be obtained from a stain or swab of a known source. For long-term storage, the specimen should be stored frozen. Specimens can be stored at refrigerated temperatures (4°C) for short periods of time prior to analysis. To maintain specimen integrity, it is critical to avoid multiple freeze-thaw conditions.

Fingernail Clippings

Fingernail clippings are also a good source of biological material for DNA analysis. The clippings collected from the victim or crime scene can be used to identify the suspect as well as to predict the sex of the attacker (Figure 2.7). In instances of a sexual assault, the victim may scratch the attacker while defending

Figure 2.6
A sample cup of urine. The epithelial cells in the urine can be concentrated by centrifugation and subjected to DNA analysis. (Courtesy of Mountainside Medical Equipment, Marcy, NY.)

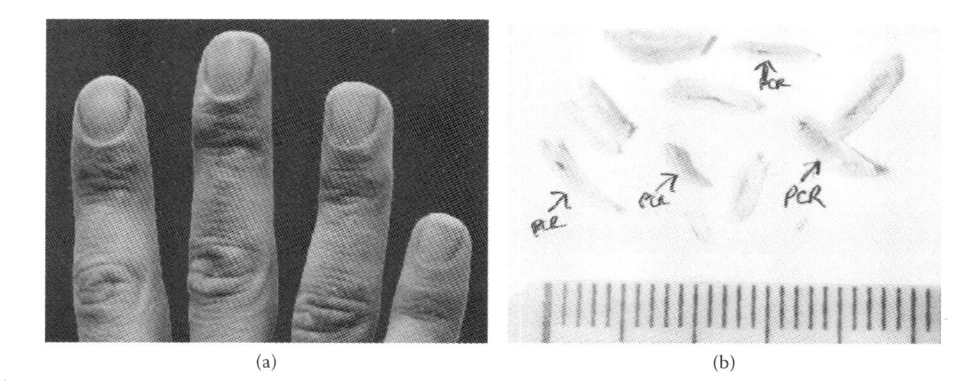

(a) (b)

Figure 2.7
Fingernail clippings. (a) Fingernail clippings can be collected from the victim and (b) prepared for DNA analysis. (Image [a] reproduced from http://www.rsc.org/Publishing/Journals/cb/Volume/2008/4/fingertips.asp. With permission from the Royal Society of Chemistry.)

him/herself and collect skin (i.e., epithelial cells) or blood underneath the fingernails. Using a sterile and moistened cotton swab, the nail clippings can be swabbed to collect the foreign material from under the nails. The swab is dried and placed in a container for storage or placed in a vial containing a small volume of sterile solution such as 1× TE buffer for short-term storage prior to DNA analysis. Swabs containing the biological material for DNA analysis are stored as recommended, as with stains.

Chapter 3

Exercise 1

Evidence Examination Using an Alternate Light Source

Introduction

A physical examination of forensic evidence thought to contain biological material using visible light may not reveal all the bodily fluids present. In fact, the composition of the evidence or the material containing the bodily fluid may determine the outcome of detecting biological material by conventional light. The use of an alternate light source (ALS) can reveal otherwise hidden or undetected biological fluids or evidence. At a crime scene or in a laboratory, fast and accurate detection of possible traces is of vital importance and is possible through the use of an ALS. Since many biological fluids fluoresce when illuminated with a specific wavelength, such "signals" can provide critical information to the investigator or laboratory analyst.

The use of ALS in forensic investigations was first used by the Royal Canadian Mounted Police in the 1970s. The ALS "units" at that time were typically water-cooled argon-ion lasers that were large and expensive machines that could not be transported to the crime scene. Instead, evidence samples were collected at the scene and brought to a laboratory for analysis. In the late 1980s, portable lasers were developed that were capable of delivering a single wavelength of light for the detection of biological fluids. Then, in the 1990s, the development of a high-intensity incandescent bulb allowed a wider range of wavelengths, from the ultraviolet range of the light spectrum to the near infrared. Currently, ALS is designed to be light and portable, and is easy to operate (Figure 3.1a).

An ALS consists of the light itself (such as a laser or incandescent bulb), a filter or combination of filters that enable all but the selected wavelengths of light to be screened out, and a device to deliver the light to the evidence being examined. If the wavelengths of light being used are potentially harmful, protective goggles equipped with a red, orange, or yellow filter are used to further screen the incoming light. The various filters allow the visualization of fluorescence (the absorption of light at one wavelength and the emission of light at a longer wavelength) emitted by the stains or biological material (Figure 3.1b).

Most organic materials can be made to fluoresce. For example, a fingerprint, invisible to the naked eye, when illuminated with blue-green light from a laser or incandescent source will fluoresce yellow. Other biological materials that fluoresce include saliva, blood, semen, urine, and vaginal secretions. Thus, the use of an ALS and glasses equipped with an orange filter will allow the crime scene investigator or laboratory analyst to observe a bright white area, in the absence of conventional light, where a stain containing biological material is present. However, given the fact that all protein-rich dried stains fluoresce, this procedure is only a presumptive test, and evidence samples are generally subjected to further testing for confirmatory analysis for biological origin. Considering requests that involve the analysis of large items

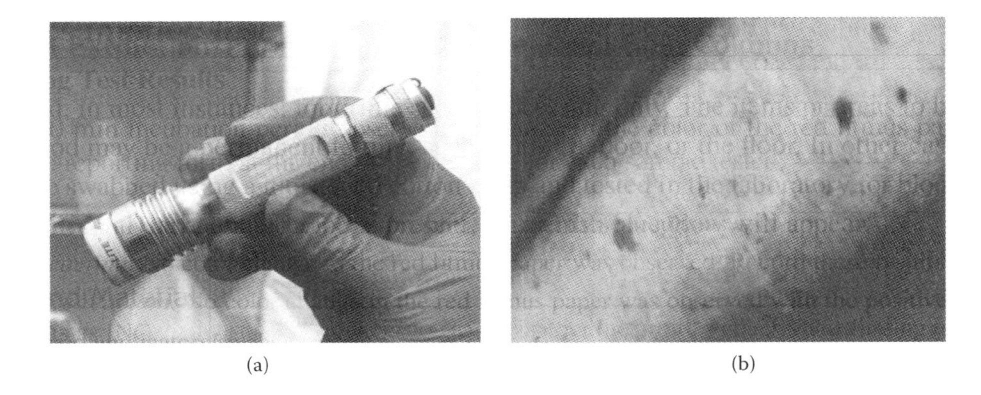

(a) (b)

Figure 3.1
The use of an ALS in the detection of bodily fluids. (a) A portable handheld alternate light source (Orion•Lite 455 nm LED Forensic Light Source). (b) A bite wound viewed using a blue ALS with an orange filter. For programs with limited resources, a blue LED flashlight and orange goggles are available for less than $200. (Courtesy of Angelia Trujillo, Forensic Nurse Examiner, University of Alaska at Anchorage.)

such as bedding items, clothing, towels, carpets, or hard surfaces, the use of ALS becomes extremely valuable in the investigative process.

Objective

The following exercise will introduce the student to the use of an alternate light source for the detection of different biological fluids on evidence collected from a crime scene. Bodily fluids found in forensic cases are saliva, blood, semen, urine, and vaginal secretions. The method introduced in this exercise is a presumptive test that will allow the student to rapidly detect and identify areas on evidence samples, based on fluoresces, that contain bodily fluids.

All biological samples should be treated as potentially infectious. The use of disposable gloves is highly recommended to prevent cross-contamination and to protect the wearer from any potential hazards associated with handling samples thought to contain bodily fluids. Goggles or shields should be worn to provide eye protection from the intense light source coming from the ALS. Appropriate sample handling and disposal techniques should be followed. A helpful organizational sheet is provided at the end of the exercise to record data, document results for any photography or diagram(s) of evidence, and note other necessary information.

Equipment and Material

1. Orion•Lite 455 nm LED Forensic Light Source (Evident, Union Hall, Virginia)
2. Orange glasses (Evident, Union Hall, Virginia)
3. Permanent marker or Sharpie
4. Bench paper
5. Disposable gloves
6. Bleach
7. Ethanol
8. Sterile 1.5 ml Eppendorf/microcentrifuge tubes

9. Sample handling tools (scissors, scalpel blades, forceps, etc.)
10. Samples containing dried saliva, blood, semen, or urine (evidentiary sample on fabric, cloth, swab, filter paper, or other nonabsorbing substrate, e.g., inanimate object)
11. Known saliva, blood, semen, or urine stains on filter paper

Procedure

1. *Preparation of site.* Select a work area or room that can be completely darkened. Once the evidence sample is illuminated with the ALS, the fluorescence can only be observed in a darkened room or area.
2. Cover the selected work area with bench paper.
3. When handling the evidence and known reference samples use proper sterile laboratory techniques. All sample handling tools should be sterilized initially and in between handling of the various samples by placing the tools first in bleach (15–30 s) and then in ethanol (15–30 s).
4. Place the ALS (Orion•Lite 455 nm LED Forensic Light Source) on the work area and make sure the unit is turned to the OFF position during the setup.
5. *Evidence examination.* Remove the items to be tested and place them on the prepared work area. Describe the item to be tested and any "stains" that might be observed with the naked eye under normal lighting conditions and record these observations in the designated forms at the end of the exercise.

Note: In a forensic laboratory the items to be tested would be labeled with a case number, the analyst's initials, and the date of the examination.

6. Place the orange filter or goggles over your eyes. Turn the lights in the work area or room OFF and turn the ALS to the ON position.
7. The light beam from the ALS will illuminate the biological fluid or stains and cause the items to fluoresce white or orange to yellow, depending on the bodily fluid being examined.
8. Using a permanent marker, circle all of the areas that fluoresce and beside each area identify each stain with either a numerical (i.e., 1, 2, 3) or alphabetical (i.e., A, B, C) code.
9. Turn the ALS to the OFF position and turn the lights to the work area or room ON.
10. Record all of your observations and descriptions of the stains that fluoresced in the designated forms at the end of the exercise.
11. *Presumptive testing.* The "marked" stains can then be removed (i.e., cut out using the sterile handling tools or "lifted" using a moistened swab) from the substrate and placed into a labeled sterile 1.5 ml microcentrifuge tube and subjected to serological testing to determine the biological origin of the fluid or stain (see Exercises 2–5).

Note: In a forensic laboratory, after the completion of the examination the evidence samples are returned to their original containers, resealed, and returned to storage following appropriate evidence handling and chain of custody procedures established by the laboratory.

Results

Interpreting Test Results

1. *Positive test results for saliva.* A dried saliva stain is virtually colorless and undetectable with the naked eye. With an excitation light with a wavelength of 450 nm (blue light) and orange filter (Figure 3.1a) or goggles the saliva stain will appear white (Figure 3.1b) (Vandenburg and Oorschot, 2006). A saliva stain will also fluoresce with a lower intensity than semen.

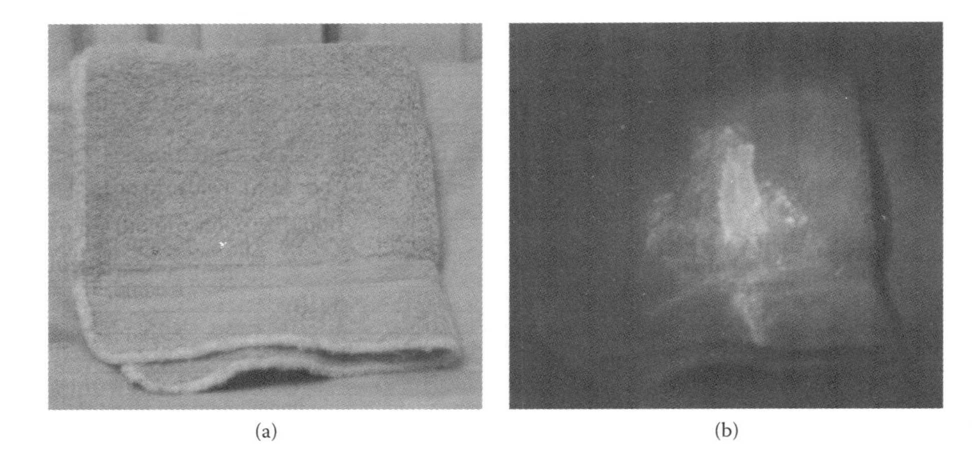

(a) (b)

Figure 3.2
Detection of semen stains on an inanimate object. (a) Under normal lighting, a semen stain is undetectable on a washcloth. (b) The same washcloth, containing the semen stain, viewed in the dark using an ALS (Surefire Blue Light and Bluemaxx Light). (Courtesy of Angelia Trujillo, Forensic Nurse Examiner, University of Alaska at Anchorage.)

2. *Positive test results for blood.* A dry bloodstain will not show a fluorescence effect; however, as a protein-rich biological fluid blood will adsorb in a very broad region of the wavelength spectrum (Stoilovic, 1991). Hence, a bloodstain will appear as a dark spot when exposed to any type of light and will appear brighter against a dark background (Virkler and Lednev, 2009; Lennard and Stoilovic, 2004).
3. *Positive test results for semen.* Under normal lighting a semen stain is undetectable (Figure 3.2a). However, a dried semen stain will have a very strong fluorescent "signal" with an ALS (Figure 3.2b). Supplying the specific excitation light (blue) with the appropriate goggles and filter (orange), a strong fluorescent (orange to yellowish) signal will be observed (Stoilovic, 1991).
4. *Positive test results for urine.* Urine stains are extremely difficult to see, especially if the fluid was deposited and became diluted on fabric. With an excitation light with a wavelength of 450 nm (blue light) and orange filter or goggles the urine stain will appear white (Vandenburg and Oorschot, 2006).

Alternate Light Source: The Detection of ________________		
Sample	Concentration	Results

Alternate Light Source Reporting Form

In the space provided above, place a photograph of the evidence that was tested or draw the area of the fabric/cloth tested using an alternate light source.

Questions

1. Why is it important to identify the biological source of the stain from evidentiary sample(s)?

2. Describe the principle of an alternative light source for the detection of bodily fluids.

3. After identifying a stain using an alternate light source, how would you collect the sample from an inanimate object (e.g., bottles, cans)? For skin?

Chapter 4

Forensic Serology

Introduction

In 1901, Karl Landsteiner published observations of agglutination of human blood cells by other human sera. Landsteiner determined that human blood was different and that blood could be distinguished by its group or type. Landsteiner's work led to the classification system, referred to as the ABO blood group system, that is used today (Landsteiner, 1962). Until the 1990s, forensic scientists used the ABO blood group system as well as other factors (e.g., Rh factor) as the best means of linking a blood sample, collected at a crime scene, to an individual or suspect. Although the ABO blood typing system was able to provide some preliminary indications about identity in criminal and paternity cases, its use was limited because there are only four possible blood types (A, B, AB, and O), and the vast majority of people in any population have two of these types: type A (41%) and type O (45%). The remaining blood types in the population are type B (12%) and type AB (3%). In other words, an individual may have a blood type consistent with the evidentiary sample tested and still not be the perpetrator of the crime or the father of the tested child. Eventually, as additional genetic markers were developed (e.g., isozymes such as phosphoglucomutase (PGM), erythrocyte acid phosphatase (EAP), and adenylate kinase (AK)), the percentage of individuals in a given population potentially linked to an evidentiary sample decreased, allowing the power of discrimination to increase (Figure 4.1).

Other biological fluids, such as saliva and semen or sperm, may require serological analysis prior to additional tests. The identification of saliva can be important in cases where oral sexual "intercourse" has been claimed or may have occurred. Additionally, the epithelial cells in saliva contain DNA that could be used for forensic DNA testing. The presence or absence of semen, as well as the identification of this fluid, can also be extremely important in sexual assault cases (Figure 4.2).

Objective

The following exercises will introduce the student to serological methods used to identify different biological fluids. The fluids found in forensic cases are saliva, blood, semen, and urine. The methods and tests introduced in these exercises are presumptive tests that are designed to (1) yield rapid test results with substrates that are convenient and available and (2) give easily interpretable results or, in some instances, a recognizable color change.

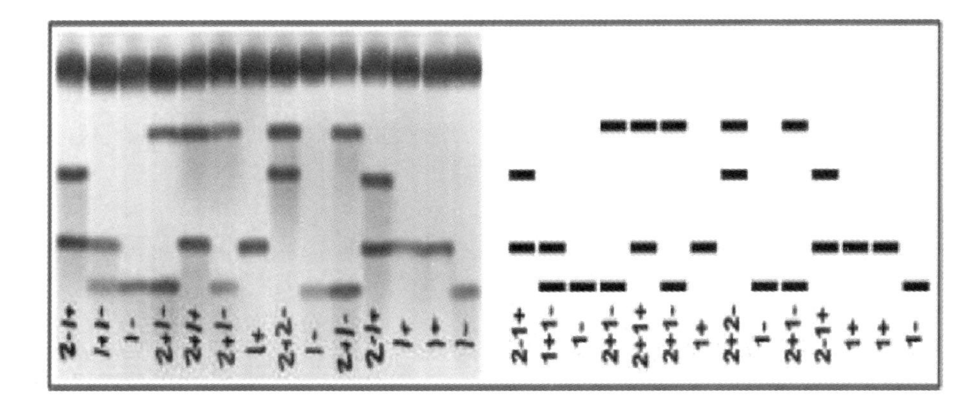

Figure 4.1
Serum protein polymorphisms. Isozymes or enzymes with different amino acid sequences that catalyze the same chemical reaction are serum proteins that display detectable polymorphisms with alleles that have frequency differences in the population to be of value in blood typing.

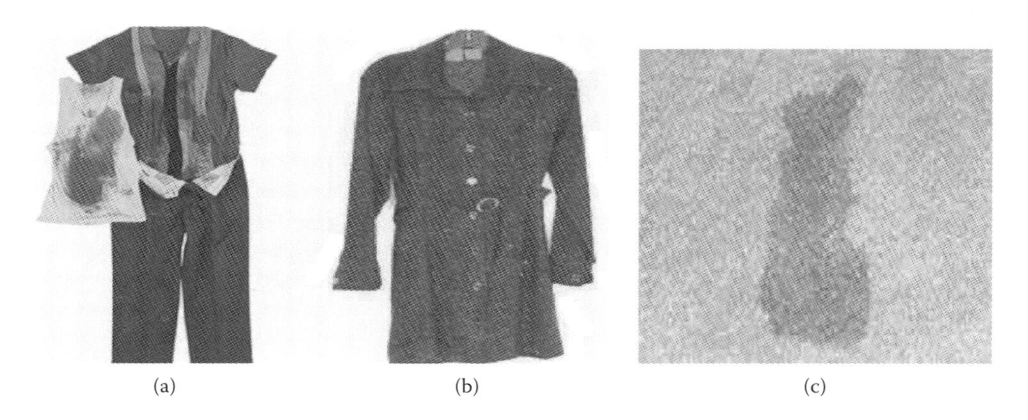

(a) (b) (c)

Figure 4.2
Examples of biological fluids collected from evidentiary samples. (a) Bloodstains on clothing that has soaked through to the inner garment of a T-shirt. (From http://www.us.ayushveda.com/remove-blood-stains-from-fabric-clothes/.) (b) Semen stain on Monica Lewinski's infamous blue dress, which almost brought down a president. (From http://www.law.umkc.edu/. . . /clinton/lewinskydress.html.) (c) A urine stain on carpet. (Courtesy of Angelia Trujillo, Forensic Nurse Examiner, University of Alaska at Anchorage.)

Chapter 5

Exercise 2
Detection of Saliva

Introduction

The identification of saliva may be important in the investigation of a criminal case where sexual assault has occurred. Saliva is composed mostly of water (98%) containing small amounts of electrolytes, buffers, glycoproteins, antibacterial compounds, and various enzymes. Saliva detection usually is based upon the presence of an enzyme known as amylase. Amylase is secreted from three pairs of human salivary glands found in the oral cavity. The process of digesting carbohydrates in the diet begins in the oral cavity when amylase from the saliva catalyzes the "breakdown" of carbohydrates, such as starch, to a variety of oligosaccharide products. Amylase is a very stable enzyme, and when detected, strongly indicates the presence of saliva.

Visual examination methods such as the use of an alternate light source (ALS) can be extremely helpful in searching for saliva stains to facilitate further analysis. The ALS consists of the light itself (such as a laser or incandescent bulb), a filter or combination of filters that enable all but the selected wavelengths of light to be screened out, a device to deliver the light to the area being examined for saliva, and appropriate protective goggles, if the wavelengths of light being used are potentially harmful, or goggles equipped with a red, orange, or yellow filter to further screen the incoming light. The various filters allow the visualization of fluorescence emitting from the stains. Although not as intense as a semen stain, the fluorescence emitted from saliva will be a presumptive test for the presence of this bodily fluid.

A demonstration of amylase activity can be used as a presumptive test to indicate the presence of saliva in an extract of a stain from either a suspect or victim, or detected even from dried articles. A very simple way to demonstrate the presence of saliva is to use the radial diffusion assay (see Figure 5.1). In this test, an agarose gel containing starch is prepared. A sample well is created by "punching" a small diameter into the agarose gel and the sample, or an extract of the sample to be tested, is placed in the well. If amylase is present, it will diffuse into the agarose gel and hydrolyze the starch. The agarose gel is subsequently stained with an iodine solution and the presence of amylase is indicated by a clear radial area in the gel away from the well. The unhydrolyzed starch in the gel will stain blue. A larger area of clear zone is more indicative of the amount of amylase present in the sample.

Objective

In this exercise you will learn to demonstrate the presence of amylase from a sample containing saliva using the radial diffusion assay. While other commercial assays are available that have been validated in forensic laboratories (i.e., rapid stain identification [RSID] test), the radial diffusion assay is simple, easy to prepare, and able to demonstrate the principles of many immunological assays used in forensic analyses (i.e., the

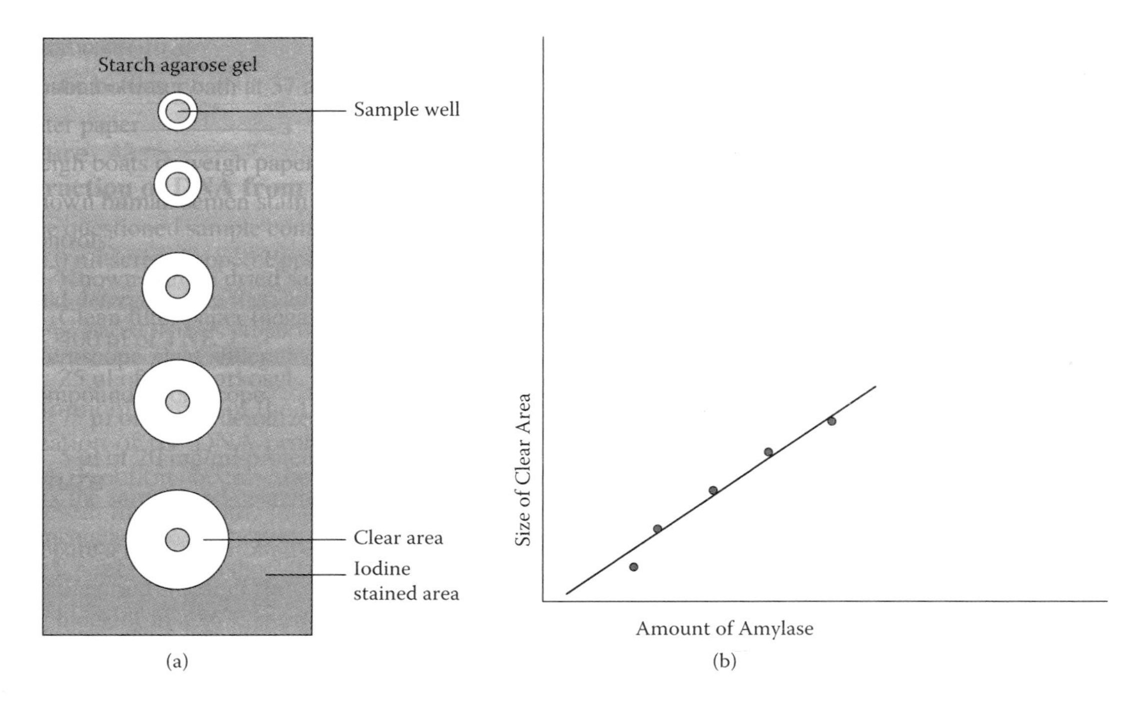

Figure 5.1
Radial diffusion assay. Amylase activity can be demonstrated using the radial diffusion assay. (a) Known amounts of amylase are applied to the wells and allowed to diffuse in an agarose gel containing starch. When an iodine solution is added to the petri plates, the radial area where starch hydrolysis has occurred will be clear and the surrounding areas will stain blue. (b) A standard curve can be used to quantify the amount of amylase. (Adapted from Farias, D.F., *Braz. J. Biol.* 70(2):405–407, 2010.)

interaction of antigens and antibodies following the diffusion of these components through a substrate to form a "line of precipitation" for identification).

Presumptive Saliva Test

The use of disposable gloves and aerosol-resistant pipet tips is highly recommended to prevent cross-contamination of samples. A helpful organizational sheet is provided at the end of the exercise to record data and other necessary information.

The Detection of Amylase in Saliva Stains

In general, a moist cotton-tipped swab is used to rub the suspected saliva stain. Forensic samples obtained from swabbing evidentiary items include envelope flaps, stamps, glass or plastic bottles, aluminum cans, and coffee lids. The swabs containing cells from saliva are placed in extraction buffer, and the final solution is analyzed in an agarose gel containing starch. Following the addition of an iodine solution, the presence of amylase is demonstrated by observing clear zones surrounding the wells. The unhydrolyzed starch in the gel will stain blue.

Equipment and Material

1. Cotton-tipped applicators/swabs
2. Extraction buffer

3. Iodine stock solution
4. Gel buffer
5. Soluble starch
6. Agarose
7. Sterile 1.5 ml Eppendorf/microcentrifuge tubes
8. Samples containing dried saliva (evidentiary sample on fabric/cloth, swab)
9. Known saliva stains on filter paper:
   a. Known human dried saliva (positive)
   b. Clean filter paper (negative)
10. Filter paper
11. Sterile scissors or scalpel
12. 90 × 90 mm petri plates
13. Gel punch or Pasteur pipet attached to an aspirator
14. Adjustable-volume digital micropipets (2–200 μl range)
15. Aerosol-resistant pipet tips
16. Disposable gloves

Procedure

1. *Preparation of sample extracts.* Remove a 1 cm^2 portion of the dried evidentiary stain from the fabric/cloth or swab and place the sample in a 1.5 ml microcentrifuge tube.
2. Add 150 μl of extraction buffer and incubate at 37°C for 1–2 h.
3. The known saliva stains or control samples should be extracted in parallel with the evidentiary samples. Remove a 1 cm^2 portion of the dried known sample from the filter paper and extract in 150 μl extraction buffer at 37°C for 1–2 h.
4. *Preparation of the starch/agarose gel.* Add 0.1 g of soluble starch and 1.0 g of agarose in 100 ml of gel buffer and heat to boiling. Boil until the agarose and starch are completely dissolved.
5. Allow the agarose-starch solution to cool to 55°C. Once the solution has cooled, pour approximately 20 ml into the 90 × 90 mm plastic petri plates. Cover the plates and allow the agarose–starch solution to solidify.
6. Using a gel punch or Pasteur pipet attached to an aspirator, create wells about 1.5 mm in diameter in the gel by removing a small plug of the solidified starch–agarose. Leave at least 1.5 cm between the holes.
7. Add 5 μl of the sample extracts to each well. Make sure to label each well with the corresponding sample. For a negative control, add 5 μl of the extraction buffer to one well.
8. Incubate the petri plates at 37°C for 16 h.

Results

Interpreting Test Results

1. *Positive test results.* A clear radial zone present around the inoculation well indicates amylase activity, and the size of the clear zone is proportional to the amount of amylase in the sample. Extracts from the known saliva stains (i.e., the positive controls) will also possess a clear, translucent radial zone around the inoculation well.

Note: A linear standard curve (in log scale) can be prepared using known concentrations of amylases (x-axis) plotted against the size of the clear radial zones (y-axis). The amount of amylase can be quantified by comparing the results from the evidentiary samples to the standard curve. Use the semilog graph paper at the end of this exercise to establish a standard curve.

2. *Negative test results.* The absence of amylase from the evidentiary samples is indicated by the absence of the clear radial zone around the well. The agarose–starch complex will appear blue up to the inoculation well. The inoculation well containing the extraction buffer (i.e., the negative control) should *not* have a clear, translucent radial zone.

   It should be noted that amylase diffusion testing is a preliminary test for amylase activity; it does not confirm the presence of saliva without additional test procedures.

Cautions: Lathia and Brendeback (1978) reported that a higher level of thiocyanate ions, as sometimes occurs in saliva from smokers, increases decolorization of the starch–iodine complex and may lead to an overestimation of amylase activity. In addition, reducing agents also reduce iodine. Prepare extracts for amylase tests using deionized water or a physiological saline solution that does not contain a reducing agent such as dithiothreitol (DTT).

The Presumptive Saliva Test: The Detection of Amylase		
Sample	Amylase Concentration	Radial Zone (cm)

Semilog Graph Paper

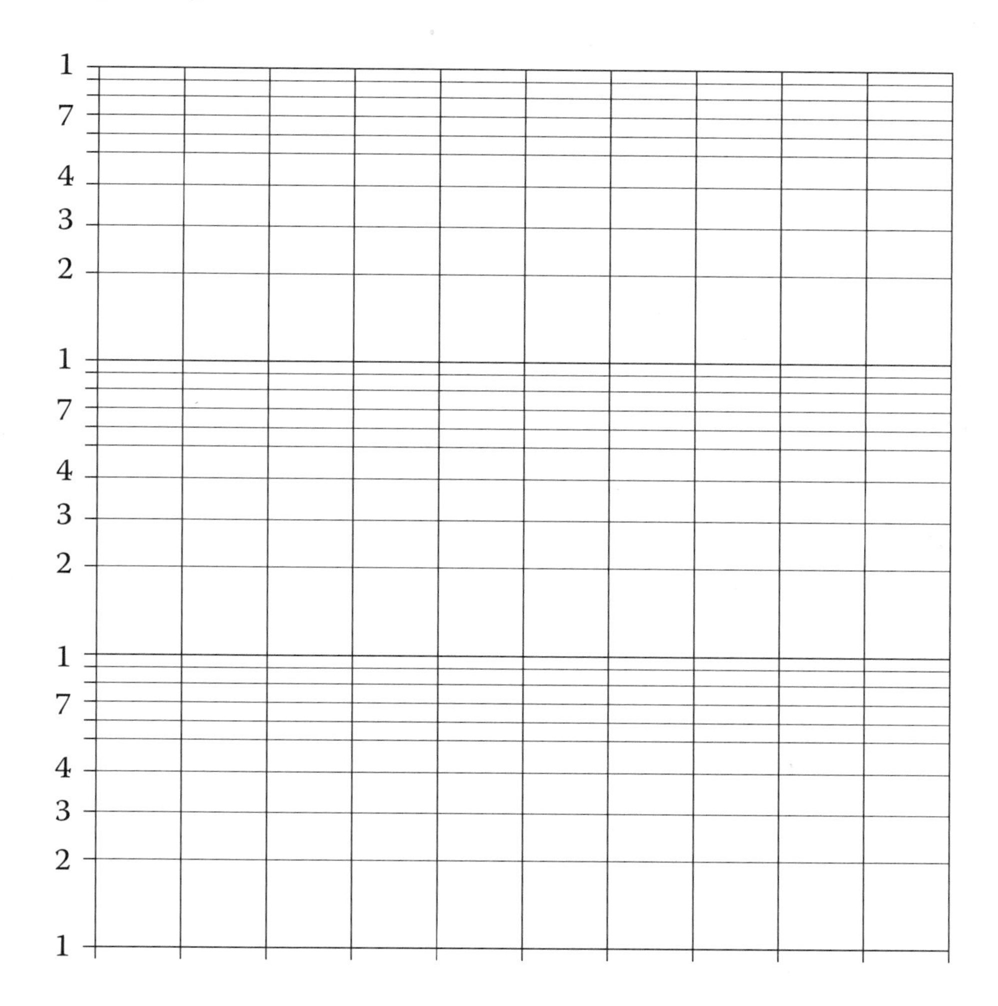

A linear standard curve (in log scale) can be prepared using known concentrations of amylases (x-axis) plotted against the size (area in cm) of the clear radial zones (y-axis). The amount of amylase can be quantified by comparing the results from the evidentiary samples to the standard curve.

Questions

1. When is it important to identify saliva in a stain or from a swab?

2. Describe the principle employed in the radial diffusion assay.

3. How would you sample an inanimate object (e.g., bottles, cans) or objects that could not be cut for the presence of amylase? For skin?

Chapter 6

Exercise 3
Detection of Blood

Introduction

Bloodstains are quite common at crime scenes where a violent crime has been committed. Using presumptive tests, the analyst must first determine if the sample is blood and if the blood is of human origin. Once the stain has been characterized as human blood, the analyst can perform confirmatory tests to generate a definitive interpretation. A number of techniques have been developed for the presumptive detection of human blood; however, a detection method that is quick and easy is necessary to determine if the stain is suitable for further analysis.

Presumptive blood tests can be performed on samples using colorless chemical reagents that form a bright color in the presence of blood and an oxidizing agent. The heme group associated with hemoglobin in red blood cells contains a peroxidase-like activity that will catalyze the oxidation or "breakdown" of certain substrates. For example, phenolphthalein, when added to a sample containing blood, will catalyze the oxidation of hydrogen peroxide and turn the mixture from colorless to a bright pink. Another example of a presumptive test based on peroxidase-like activity of hemoglobin is the leucomalachite green (LCG) test, where the reduced substrate (LCG) is oxidized from a green color to a blue-green product.

Another presumptive test for the presence of blood is the use of luminol, a chemical that glows greenish blue when it reacts with blood. Luminol can detect highly degraded blood or minute samples. Investigators typically use luminol at crime scenes to reveal bloodstains that have been washed out, wiped off, or not visible to the naked eye. The chemical is sprayed over the area to reveal blood, and the chemiluminescence is observed in total darkness. Luminol has been reported to be so sensitive that it can detect as little as 0.2–10 parts per million, or (on average) 1 drop of blood within a container of 999,999 drops of water. Photographs are taken to obtain a permanent record of the test results.

Objective

In this exercise you will learn three different presumptive techniques used in forensic serology to determine the presence of blood in samples. These techniques are representative examples of modern tests or assays used in forensic laboratories to detect the presence or absence of human blood from known samples, as well as from evidentiary samples. The presumptive blood tests described below are relatively brief and easy to perform. When completed, the remaining sample can be disposed of appropriately in a biohazard container or utilized in the isolation of DNA in the subsequent exercises.

Presumptive Blood Tests

The use of disposable gloves and aerosol-resistant pipet tips is highly recommended to prevent cross-contamination. A helpful organizational sheet is provided at the end of the exercise to record data and other necessary information.

Phenolphthalein Presumptive Test

In general, a moist cotton-tipped swab is used to rub the suspected stain. A phenolphthalein solution is placed on the cotton swab followed by the addition of hydrogen peroxide. If blood is present, a bright pink color will appear within 10–15 s of the addition of the phenolphthalein solution.

Equipment and Material

1. Cotton-tipped applicators/swabs
2. 95–100% ethanol
3. Phenolphthalein stock solution
4. 3% hydrogen peroxide
5. Samples containing dried blood (evidentiary sample on fabric/cloth or FTA card)
6. Controls:
   a. Known human dried bloodstain (positive)
   b. Clean filter paper (negative)
7. Filter paper
8. Sterile scissors or scalpel
9. Lancet or fingerstick device

Procedure

1. *Fingerstick method.* Draw a circle on the fabric/cloth to indicate a collection area. Place either the cloth or FTA card so that the filter paper side containing the drawn or printed circles is facing up. Before any fingerstick is attempted, ensure that the area to be pricked on the individual's finger is cleansed thoroughly with an alcohol swab. Prick any finger except the right index finger, using the fingerstick device provided in the collection supplies. Each fingerstick device is designed to be used only once for safety purposes. Squeeze the pricked finger to produce droplets of blood. Touch the blood droplets to the appropriate circled areas of the fabric/cloth or the FTA card so that the circled areas are at least half covered with blood. If more than one fingerstick is necessary, use a different finger for each stick until both circles are at least 50% covered with blood.
2. Using a pair of scissors or a sterile scalpel, place a small cutting of the dried bloodstain (the evidentiary sample from the fabric/cloth or FTA card) on a piece of clean filter paper.

   OR

   Gently rub a suspected stain with a sterile cotton swab moistened with sterile distilled water and place on a piece of clean filter paper.
3. Also place the positive and negative controls on a piece of filter paper.
4. Add one drop of 95–100% ethanol to each stain or swab.
5. Add one drop of phenolphthalein to each stain or swab.
6. Add one drop of 3% hydrogen peroxide to the stains or swabs.
7. Observe the color change. The positive control should turn bright pink immediately after adding the hydrogen peroxide. The negative control should remain colorless after the addition of each chemical.

Note: The positive and negative controls must work properly in order for the results from the evidentiary samples to be credible. A false positive is indicated when a bright pink color does not form immediately after the addition of the hydrogen peroxide.

8. Record the blood screening test results (positive, negative, or inconclusive) in the table provided.

Leucomalachite Green (LCG) Presumptive Test

A moistened cotton-tipped swab is used to rub the suspected stain. A drop of the LCG solution is placed on the swab followed by the addition of hydrogen peroxide. If blood is present, a bluish green color will immediately appear.

Equipment and Material

1. Cotton-tipped applicators/swabs
2. LCG stock solution
3. 3% hydrogen peroxide
4. Samples containing dried blood (evidentiary sample on fabric/cloth or FTA card)
5. Controls:
   a. Known human dried bloodstain (positive)
   b. Clean filter paper (negative)
6. Filter paper
7. Sterile scissors or scalpel
8. Lancets or fingerstick device

Procedure

1. *Fingerstick method.* Draw a circle on the fabric/cloth to indicate a collection area. Place either the cloth or FTA card so that the filter paper side containing the drawn or printed circles is facing up. Before any fingerstick is attempted, ensure that the area to be pricked on the individual's finger is cleansed thoroughly with an alcohol swab. Prick any finger except the right index finger, using the fingerstick device provided in the collection supplies. Each fingerstick device is designed to be used only once for safety purposes. Squeeze the pricked finger to produce droplets of blood. Touch the blood droplets to the appropriate circled areas of the fabric/cloth or the FTA card so that the circled areas are at least half covered with blood. If more than one fingerstick is necessary, use a different finger for each stick until both circles are at least 50% covered with blood.
2. Moisten a cotton swab with distilled water and gently rub over the sample containing the suspected bloodstain. If the bloodstain is on clothing, cut a small (no more than 0.2 mm^2) portion of the stain from the cloth.
3. Set up a positive and a negative control. The positive control should be known blood spotted onto a swab, piece of cloth, or FTA card. The negative control should be a clean swab, piece of cloth, or no substrate.
4. Apply one drop of the LCG stock solution to the swab or cutting.
5. Apply one drop of 3% hydrogen peroxide to the swab or cutting.
6. Observe the color change. Immediate appearance of bluish green color is indicative of the presence of blood.
7. The positive control should turn blue-green immediately after adding the hydrogen peroxide. The negative control should remain colorless after the addition of each chemical.

Note: The positive and negative controls must work properly in order for the results from the evidentiary samples to be credible. A false positive is indicated when a blue-green color does not form immediately after the addition of the hydrogen peroxide.

8. Record the blood screening test results (positive, negative, or inconclusive) in the table provided.

Luminol Presumptive Test

Luminol is used, in most instances, for crime scene processing only. The items or areas to be tested for the presence of blood may be a permanent fixture, such as a wall, door, or the floor. In other cases, the items to be tested can be swabbed using a moistened cotton swab and tested in the laboratory for blood. The luminol is applied, and in total darkness, if blood is present, a greenish blue glow will appear.

Equipment and Material

1. Cotton-tipped applicators/swabs
2. Luminol solution (prepared in an atomizer just prior to use)
3. Known human bloodstain treated with a solution of detergent in water or dilute HCl (evidentiary sample on fabric/cloth or FTA card)
4. Controls:
   a. Known human dried bloodstain (positive)
   b. Clean filter paper (negative)
5. False positives:
   a. Known dried bleach (prepared fresh) on filter paper or cloth
   b. Copper source
6. Filter paper
7. Sterile scissors or scalpel
8. Lancet or fingerstick device

Procedure

1. *Fingerstick method.* Draw a circle on the fabric/cloth to indicate a collection area. Place either the cloth or FTA card so that the filter paper side containing the drawn or printed circles is facing up. Before any fingerstick is attempted, ensure that the area to be pricked on the individual's finger is cleansed thoroughly with an alcohol swab. Prick any finger except the right index finger, using the fingerstick device provided in the collection supplies. Each fingerstick device is designed to be used only once for safety purposes. Squeeze the pricked finger to produce droplets of blood. Touch the blood droplets to the appropriate circled areas of the fabric/cloth or the FTA card so that the circled areas are at least half covered with blood. If more than one fingerstick is necessary, use a different finger for each stick until both circles are at least 50% covered with blood.
2. Using a pair of scissors or a sterile scalpel, place a cutting of the dried bloodstain (the evidentiary sample from the cloth/fabric or FTA card) on a piece of clean filter paper.

   OR

   Gently rub a suspected stain with a sterile cotton swab moistened with sterile distilled water and place on a piece of clean filter paper.
3. Draw the area on the fabric/cloth or area that is to be tested in the section provided at the end of the exercise.
4. In the darkness, apply (or spray) the area to be tested evenly with luminol. Be careful *not* to oversaturate the area.
5. Observe the tested area and the area(s) that produce chemiluminescence or the greenish blue glow. Record the results in the reporting form at the end of the exercise.

Results

Interpreting Test Results

Phenolphthalein Presumptive Test

1. *Positive reaction.* An immediate pink color develops if blood is present. Remember, this is not a confirmatory test for blood.
2. *Negative reaction.* No color change or the development of a color occurs.
3. *Inconclusive reaction.* A color develops other than that described above for the positive reaction.
4. *False-positive reactions.* Chemical oxidants and catalysts (e.g., copper and nickel salts, rust, and formalin) may produce a color change before the addition of hydrogen peroxide. Peroxidases from certain plants (e.g., cabbage, horseradish, celery, and carrots) can often give a delayed color reaction.
5. The controls must work in order to report the test results.

Luminol Presumptive Test

1. *Positive reaction.* A greenish blue light will be omitted (chemiluminescence) in the dark and in the presence of blood and will last from a few seconds to several minutes.

   Wipe marks, fingerprints, footprints (Figure 6.1), drag marks, and spatter patterns are often visualized at crime scenes when spraying luminol on an area containing blood not visible to the naked eye. Remember, this is not a confirmatory test for blood.
2. *Negative reaction.* No light will be omitted or visualized from the test area in the dark.
3. *Inconclusive reaction.* If all items tested or sprayed with luminol appear to omit light (chemiluminescence), the substrate is reacting with the chemical and the test results should be reported as inconclusive.
4. *False-positive reaction.* A light will be omitted but will appear as sparkles and not a constant light being omitted.

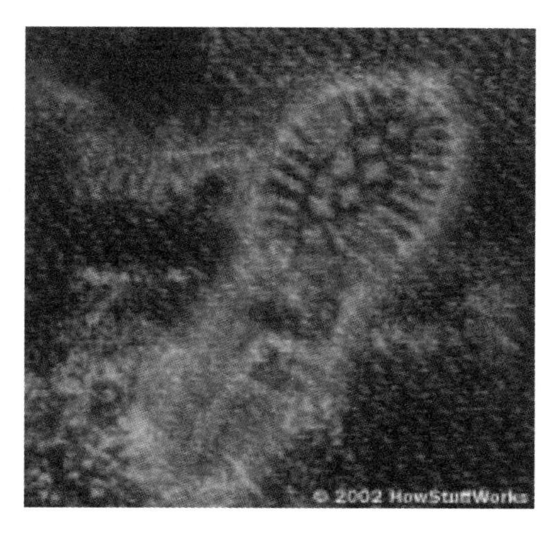

Figure 6.1
Luminol enhancement at a crime scene. Crime scene investigators will typically apply luminol on areas where no blood is visible. In the presence of blood luminol will react and emit a greenish blue light as seen above. The glow that is detected is photographed and documented by the investigator. (From Harris, T., How luminol works, June 11, 2002, http://science.howstuffworks.com/luminol.htm (accessed September 28, 2012).)

Phenolphthalein Presumptive Test

Analyst: ______________________ Lab number: ______________________

Date: ______________________

+ Positive for the presence of blood

– Negative for the presence of blood

INC Inconclusive

Comments: ______________________

Item Number/Description	Results	Item Number/Description	Results

Leucomalachite Green Presumptive Test

Analyst: ______________________ Lab number: ______________________

Date: ______________________

+ Positive for the presence of blood

– Negative for the presence of blood

INC Inconclusive

Comments: ______________________

Item Number/Description	Results	Item Number/Description	Results

Luminol Presumptive Test Reporting Form

In the space provided above draw the area of the fabric/cloth or the area to be tested for the presence of blood using luminol.

Questions

1. What is the purpose of the positive and negative controls in the phenolphthalein presumptive test?

2. If the results of a test are positive and the test was performed on a stain from a substrate (i.e., fabric) why is it important to test the adjacent material next to the stain, in other words, perform a substrate control test?

3. What would be some potential drawbacks from the use of luminol at a crime scene?

Chapter 7

Exercise 4

Detection and Identification of Semen

Introduction

Human semen (seminal fluid) usually contains spermatozoa or sperm cells and the combined secretion of several accessory glands. In humans, seminal fluid contains several components besides spermatozoa, such as proteins (P30, prostate specific antigen, and flavin) and enzymes (acid phosphatase), which can be exploited for the presumptive and confirmatory identification of this body fluid. Visual examination with an alternate light source (ALS) will facilitate the detection of semen stains due to the flavin from the seminal fluid, which will fluoresce under ultraviolet light. Other presumptive tests exist for the detection of semen that are based on the presence of acid phosphatase (AP). However, since AP activity is not exclusive to human semen, confirmatory assays are necessary for the positive identification of semen.

Confirmatory tests for the detection and identification of semen can be performed by staining the suspected semen extracts, followed by microscopic examination of the specimen(s). In this procedure, a portion or a section of the suspected semen stain is cut from the fabric and placed in water. The fabric containing the cells is gently vortexed, and the extract, now containing the cells, can then be transferred to a microscope slide. Alternatively, the suspected stain on the fabric can be dampened and rubbed onto the microscope slide for staining and observations.

Microscopic observations of sperm cells provide strong evidence of the presence of seminal fluid. To facilitate the detection and identification of sperm cells, stains or dyes are often used to aid in the microscopic examination of the stain extract. The most common staining technique used to detect the presence of human sperm is the Christmas tree stain (Figure 7.1) (Allery et al., 2001). In this staining process two dyes are used: nuclear fast red and picroindigocarmine (PIC). The red component or nuclear fast red will stain the nuclei of sperm and other cells red, whereas PIC will stain the neck and tail portions of the sperm and epithelial cells green.

Objective

In this exercise you will learn to demonstrate the presence of semen from a sample containing a stain thought to contain semen using the acid phosphatase (AP) test and the Christmas tree stain. While other commercial assays are available, both tests, which have been validated in forensic laboratories, are used for the detection of seminal fluid or semen. The Christmas tree stain detects sperm, while the AP test, as well as many other assays, demonstrate the presumptive presence of semen. While both tests are simple and easy to prepare, the Christmas tree stain is able to demonstrate the principles of many staining assays used in forensic analyses for detection and identification of bodily fluids.

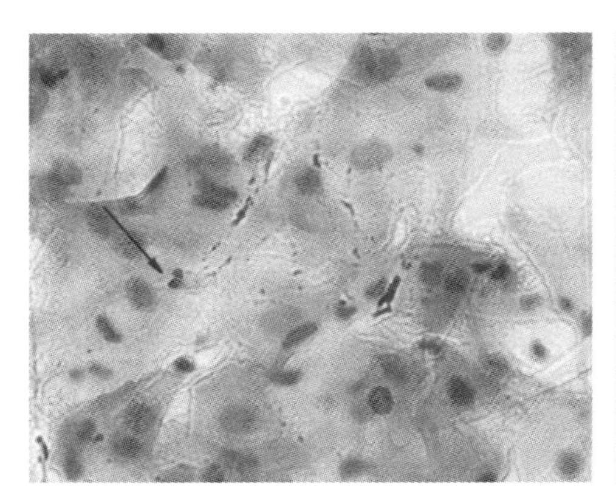
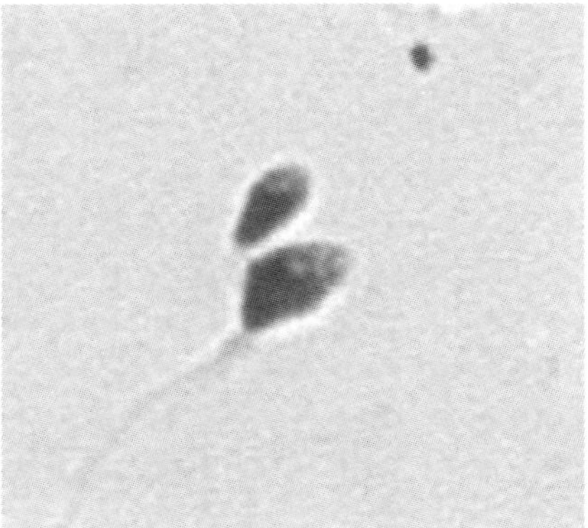

Figure 7.1
A Christmas tree stain of a vaginal smear from a sexual assault victim. The red component or the nuclear fast red stain will stain the nuclei of sperm (see arrow on left) and other cells red, whereas the green component of the PIC will stain the neck and tail portions of the sperm and epithelial cells green (background color).

The use of disposable gloves, protective eyewear, and aerosol-resistant pipet tips is highly recommended to prevent cross-contamination. A helpful organizational sheet is provided at the end of the exercise to record data and other necessary information.

The Acid Phosphatase (AP) Test

The AP test is a presumptive test for the presence of seminal fluid or semen. Acid phosphatase is an enzyme that is secreted by the prostate gland and is found in large amounts in seminal fluid. Since the enzyme is found in other biological material, the AP test is only a presumptive test that can be verified with other confirmatory tests.

The AP test is a colorimetric test whereby acid phosphatase cleaves the phosphate from α-naphthyl phosphate to form naphthol. Naphthol and buffered Brentamine Fast Blue B form a deep purple azo dye, and hence a color reaction indicating the presence of seminal fluid.

Equipment and Material

1. 10% bleach
2. Sterile distilled water
3. 100% ethanol
4. AP spot test solution (Serological Research Institute, Richmond, California)* [see p. 37]
5. Swabs
6. Sample handling tools (e.g., scissors, forceps, and scalpel blades)
7. Clear test tubes (6 × 50 mm)
8. Known human semen stain (evidentiary sample on a fabric)
9. Positive controls (known human dried sample)
10. Negative control (filter paper containing all reagents but no sample)
11. Filter paper or absorbent pad

Procedure

1. Take a small cutting of the question sample and place on filter paper/absorbent pad or swab the questioned evidence stain. Use a known semen stain as a positive control and an unstained swab or filter paper as a negative control.

* **Note:** If the AP spot test solution is purchased from a commercial supplier, follow product insert regarding preparation and storage requirements. If the solution is being manually prepared, follow the directions under "AP spot test solution" in Appendix A.

2. Add one or several drops (depending on the size of the sample) of the AP spot test solution, if prepared from a commercial source, to both the known positive and negative controls and the evidentiary sample. If the AP test solution is manually prepared, add one drop of solution A followed by one drop of solution B.
3. Observe the tested area for approximately 1 min for a color change.

Results

Interpreting Test Results

1. *Positive reaction.* A purple color change indicates the presence of seminal fluid, which is a positive test result.
2. *False positives.* AP is found in other bodily fluids. False-positive results may occur due to vaginal secretions (bacterial infection and pregnancy cause increased levels of AP in the vaginal area), whereby a color reaction is slow and faint; in fecal stains the color reaction is slow and faint and, in blue dye, transfers from fabric or other objects. Plant matter, spermicides, and some feminine hygiene products can cause a false-positive test result.
3. *Negative reaction.* No color change or pink color change in the tested area after 1 min indicates the absence of seminal fluid and is a negative result.
4. The controls must work in order to report the test results.
5. Record your test results in the form provided at the end of this exercise.

The Christmas Tree Stain

Unlike the AP test, the Christmas tree stain can detect the presence of sperm. Using Kernechtrot and Picroindigocarmine staining solution, sperm heads from vaginal swabs can be stained and detected by microscopy.

Equipment and Material

1. Kernechtrot staining (KS) solution
2. Picroindigocarmine staining (PICS) solution
3. Flame or heat block
4. Distilled water
5. 95% ethanol or methanol
6. Filtration apparatus
7. 500 ml glass beakers
8. Balance
9. Spatula

10. Glass rod
11. Plastic bottles
12. Filter paper
13. Weigh boats or weigh paper
14. Known human semen stain (evidentiary sample on a fabric)
15. Controls:
    a. Known human dried sample (positive)
    b. Clean filter paper (negative)
16. Microscope glass slides
17. Compound microscope

Procedure

1. Prepare a thin smear of an extract of a suspected and known semen stain on a glass slide and allow it to air dry.
2. Fix the dried smear of the extract by gently "flaming" the glass slide over a flame or incubate the slide on a heating block or in a 56°C oven for 30 min.
3. Add two to five drops of the KS reagent (red reagent) to cover the fixed smear on the microscope slide and allow it to stand at room temperature for 15 min.
4. Wash off the slide with a gentle stream of distilled water and drain the slide.
5. Add two to five drops of the PICS reagent (green reagent) to cover the stained smear on the microscope slide and allow it to stand at room temperature for no more than 5–15 s.
6. Wash off the PICS reagent with 95% ethanol or methanol and dry the slide at room temperature.
7. Examine the stain immediately at 400× magnification using a compound microscope.

Results

Interpreting Test Results

1. *Positive reaction.* The cytoplasm of the epithelial cell or skin cells will stain green with red nuclei. If present, the sperm will stain red with green tails. The sperm head stains differentially such that the cap stains pink and the sperm base stains red.
2. *Negative reaction.* The cytoplasm of the epithelial cells will stain green with red nuclei. Differentially stained sperm will not be observed.
3. The controls (i.e., positive and negative) must work in order to report the test results.

Acid Phosphatase Test Reporting Form

Analyst: ______________________ Lab number: ______________________

Date: ______________________

+	Positive for the presence of seminal fluid or semen
–	Negative for the presence of seminal fluid or semen
INC	Inconclusive

Comments: ______________________

____

____

Item Number/Description	Results	Item Number/Description	Results

Christmas Tree Test Reporting Form

Analyst: ____________________ Lab number: ____________________

Date: ____________________

+ Positive for the presence of seminal fluid or semen

– Negative for the presence of seminal fluid or semen

INC Inconclusive

Comments: ____________________

____

____

Item Number/Description	Results	Item Number/Description	Results

Christmas Tree Stain Reporting Form

In the space provided above draw the image that is observed under the microscope of the stained smear on the slide. Label the epithelial and sperm cells.

Questions

1. Why is it important to determine the presence or absence of sperm in evidentiary samples?

2. Describe a positive and a negative reaction in the AP test. What does a pink color stain indicate when a sample is tested for the presence of seminal fluid?

3. A young pregnant female presents to the emergency room and claims to have been raped by her boyfriend. During the examination, the victim is diagnosed with a urinary tract infection. Vaginal swabs are collected and sent to the laboratory for serological analysis. As the analyst, what tests would you perform (and why) to determine the presence of semen?

4. What is the purpose of heat fixing the dried smear to the microscope slide?

5. Microscopic observations at 400× magnification of the suspected stained sample reveal numerous red nuclei. How would you determine or differentiate red sperm heads from the red nuclei of epithelial cells?

Chapter 8

Exercise 5

Detection and Identification of Urine

Introduction

When clothing or an evidentiary sample is thought to contain a urine stain the detection of the typical yellow color, if on a lightly colored fabric, and a characteristic urine odor may be helpful. In instances where no stains are visible, the use of an alternate light source (ALS) may be helpful in locating the stain. In general, an ALS consists of a light source (such as a laser or incandescent bulb) and a filter or combination of filters that enable all but the selected wavelengths of light to be screened out that delivers the light to the evidentiary sample being examined. Using appropriate viewing accessories (such as protective goggles, if the wavelengths of light being used are potentially harmful, or goggles equipped with a filter to screen out the harmful wavelengths), urine as well as some other bodily fluids will absorb the light and fluoresce in the dark. The identified area on the sample can then be swabbed and tested for the presence of urine (see Exercise 1).

Several simple and rapid presumptive tests for the detection of urine are available. In 1886, Jaffe, a German biochemist, developed a presumptive test for the detection of creatinine (a waste product) that is secreted in urine (Jaffe, 1886; Gaensslen, 1983). The test is based on the reaction of picric acid with creatinine and a weak base to form a deep orange-red color. In 1948, Cook developed a test for urine that detects urea using an enzyme called urease. In the presence of urease, urea is cleaved to form ammonia, which is detected using bromothymol blue, an acid-base indicator. The test is conducted using litmus paper, and if a positive reaction occurs, the paper will turn from red to a bright blue color. Most forensic laboratories use either the Jaffe color test for creatinine or the urease test for urea as a presumptive screen. To confirm the presence of urine, which will consume an additional sample, a test kit (BFID-Urine) is commercially available through Independent Forensics, Hillside, Illinois.

Objective

In this exercise you will learn two different identification techniques used in serology to detect the presence of urine in evidentiary samples. The techniques used in this exercise are representative examples of modern techniques used in forensic laboratories to identify urine from stains on clothing or samples thought to contain urine. The procedures described below are relatively brief and easy to perform. Upon completion of this exercise the student will be able to state the presumptive presence of urine on the material examined, and in some instances the remaining sample can be utilized in the isolation of DNA in the subsequent exercises.

Presumptive Tests for Urine

The use of disposable gloves and aerosol-resistant pipet tips is highly recommended to prevent cross-contamination. A helpful organizational sheet is provided at the end of the exercise to record data and other necessary information.

Creatinine Presumptive Test (the Jaffe Reaction)

In general, a piece of fabric presumed to contain urine or a moist cotton-tipped swab is used to rub the suspected stain and placed in a "spot plate" along with picric acid and a weak base. In the presence of creatinine, which would indicate the presence of urine, a deep orange color will immediately develop.

Equipment and Material

1. Saturated picric acid
2. 5% sodium hydroxide
3. 95–100% ethanol
4. Sterile scissors or scalpel
5. Sterile tweezers
6. Sterile distilled water
7. Disposable pipets
8. White spot plates
9. Disposable gloves
10. Known human dried urine stain on a piece of fabric (positive control)
11. Substrate control of material or piece of fabric (negative control)
12. Reagent blank—picric acid and sodium hydroxide (negative control)

Procedure

1. From the fabric containing the suspected urine stain, cut out approximately a 2 cm^2 piece of material and set aside. At the same time, using scissors or a scalpel that have been sterilized with 95–100% ethanol and rinsed with sterile water, cut out approximately a 2 cm^2 piece of material from both the positive and negative controls and set aside.
2. Using sterile tweezers, place the 2 cm^2 cuttings of fabric from the suspected stain and the controls into separate wells of a white spot plate. Use the white spot plate reporting form at the end of the exercise to record the location of each piece of fabric for each well and the test results.
3. To each well, add one drop of picric acid and make sure each sample is completely saturated.
4. To each well, add one drop of sodium hydroxide.

Note: The concentration of sodium hydroxide may need to be adjusted as necessary to ensure that the positive and negative controls yielded expected results prior to testing the evidentiary samples.

Results

Interpreting Test Results

1. *Positive reaction.* The solution and the material inside the well will immediately form a deep orange color.
2. *Negative reaction.* The solution and the material inside the well will not react, and thus there will be no color change.

3. *Inconclusive reaction.* A slight orange color develops; compare these results to the negative control.
4. The positive and negative controls must work in order to report the test results.

Urease Presumptive Test

In general, a piece of fabric presumed to contain urine or a moist cotton-tipped swab is used to rub the suspected stain and is placed in a test tube along with water, the urease reagent, and litmus paper. In the presence of urea, ammonia is formed, which is detected using bromothymol blue, an acid-base indicator. The test is conducted using litmus paper, and if a positive reaction occurs, the litmus paper will turn from red to a bright blue color.

Equipment and Material

1. Urease buffer reagent (Sigma Aldrich)
2. 95–100% ethanol
3. Red litmus paper
4. Sterile scissors or scalpel
5. Sterile tweezers
6. Sterile distilled water
7. Disposable pipets
8. Heat block or incubator (37°C)
9. Disposable gloves
10. Test tubes (10 × 75 mm)
11. Cork stoppers (to seal the test tubes)
12. Known human dried urine stain on a piece of fabric (positive control)
13. Substrate control of material or piece of fabric (negative control)
14. Reagent blank—picric acid and sodium hydroxide (negative control)

Procedure

1. From the fabric containing the suspected urine stain, cut out approximately a 2 cm^2 piece of material and set aside. At the same time, using scissors or a scalpel that has been sterilized with 95–100% ethanol and rinsed with sterile water, cut out approximately a 2 cm^2 piece of material from both the positive and negative controls and set aside.
2. Using sterile tweezers, place the 2 cm^2 cuttings of fabric from the suspected stain and the controls into appropriately labeled test tubes. Use the test tubes reporting form at the end of the exercise to record the location of each piece of fabric for each test tube and the test results.
3. Add three to four drops of distilled water to each tube.
4. Add six to seven drops of urease buffer reagent to each tube.
5. Using a scalpel cut a slit at the small end of the cork stoppers. Prepare cork stoppers for each test tube containing a sample to be tested. Insert a strip of red litmus paper into each slit at the end of the cork stopper.
6. Place the cork stopper, containing the red litmus paper, into each test tube. The litmus paper should not touch the liquid in the tube.
7. Incubate the samples at 37°C for 30 min.

Results

Interpreting Test Results

After the 30 min incubation period observe any change in the color of the red litmus paper. Record the results using the reporting form at the end of the exercise for each sample tested.

1. *Positive reaction.* The red litmus paper turns blue. Record these results as +.
2. *Negative reaction.* No color change to the red litmus paper was observed. Record these results as –.
3. *Inconclusive reaction.* No color change in the red litmus paper was observed with the positive control. Record these results as INC.

Note: Inconclusive test results may be due to insufficient quantities of urine or that the controls used in the testing did not meet the quality control requirements.

4. The positive and negative controls must work in order to report the test results.

White Spot Plate Reporting Form
Well Contents
Well 1
Well 2
Well 3
Well 4
Well 5
Well 6
Well 7
Well 8
Well 9
Well 10
Well 11
Well 12
Plate prepared by:
Date:

Test Tubes Reporting Form	
Test Tube Contents	Results
Test tube 1	
Test tube 2	
Test tube 3	
Test tube 4	
Test tube 5	
Test tube 6	
Test tube 7	
Test tube 8	
Test tube 9	
Test tube 10	
Test tube 11	
Test tube 12	
Plate prepared by:	
Date:	

Questions

1. Why is it important to determine the presence or absence of urine from an evidentiary sample?

2. If a stain from an evidentiary sample tests positive for the presence of urine, the sample could be subjected to DNA testing. What would be the source of the DNA in this type of sample?

Chapter 9

Exercise 6
DNA Extraction

Introduction

There are a number of different approaches for the isolation of genomic DNA. Each procedure begins with some form of cellular lysis, followed by deproteinization and recovery of DNA. The main differences between the various approaches lie in the extent of deproteinization and the size of the DNA isolated. In addition, the isolation or extraction of DNA will vary according to the type of biological sample, the amount of evidence or biological sample, and the type of cell(s) present in the sample.

DNA must first be separated from the rest of the cellular components, as well as from any nonbiological material present. The removal of extraneous substances following cell lysis minimizes sample (DNA) degradation due to cellular enzymes while ensuring maximum enzymatic efficiency during the typing procedure.

Objective

In this exercise you will learn different isolation techniques used in forensic DNA analysis to extract DNA from various biological sources. The techniques used in this exercise are representative examples of modern techniques used in forensic laboratories to isolate whole genomic DNA from known samples, as well as from evidentiary samples. The extraction procedures described below are relatively brief and easy to perform. The optimum isolation procedure is highlighted for each sample or cell type. When completed, the DNA isolated in this exercise can be utilized in the subsequent exercises.

Extraction Methods

The use of disposable gloves and aerosol-resistant pipet tips is highly recommended to prevent cross-contamination. A helpful organizational sheet is provided at the end of the exercise to record data and other necessary information.

Chelex Extraction

When a minimal amount of sample is available (i.e., spot of blood), the Chelex extraction method is used. The sample is boiled in a solution containing minute beads of a chemical called Chelex. The boiling causes the cells to lyse, releasing the DNA. The Chelex binds to the extraneous cellular material, and the entire "complex" is removed by centrifugation, leaving the DNA in the supernatant. Since the high temperatures disrupt the two strands of the DNA, generating single-stranded molecules, this extraction process is generally reserved for polymerase chain reaction (PCR)-based typing techniques.

Equipment and Material

1. 15 ml sterile polypropylene test tube
2. Sterile 1.5 ml Eppendorf/microcentrifuge tubes
3. 5 ml pipettor with sterile tips
4. 1,000 μl micropipettor with sterile tips
5. 10% suspension of Chelex resin beads
6. 100 μl micropipettor with sterile tips
7. 1.5 ml test tube holder
8. 1× phosphate-buffered saline (PBS)
9. Disposable gloves
10. Boiling water bath in a 1,000 ml beaker
11. Ice in buckets
12. Table top clinical centrifuge
13. Microcentrifuge
14. Sterile cotton-tipped applicators

Procedure

Collection of Cells (e.g., buccal cells, liquid blood, cultured human cells)

1. Label a 15 ml polypropylene test tube and the top of a 1.5 ml Eppendorf tube (also referred to as a microcentrifuge tube) with your name and any other appropriate information.
2. Pipet 10 ml of suspended cells (maximum 5×10^6 cells/ml) or liquid sample into the polypropylene test tube (for harvesting cultured human cells, see the "Salting Out" procedure below, steps 1–6). For buccal cells, rinse your mouth with 10 ml of 1× PBS solution and vigorously swish against your cheeks for 10 s. Expel the PBS solution back into the labeled 15 ml polypropylene test tube over the sink.

   OR

   If sterile swabs are available, place the swab inside your mouth and press it firmly against the inside of your cheek. Roll the swab back and forth over the inside surface of your cheek at least 10 times. Repeat on the other cheek. Place the swab into a labeled 15 ml test tube containing 2 ml of 1× PBS solution. After gentle swirling the cells will dislodge from the swab in 10–15 min.

Concentrate Cells by Centrifugation

3. Centrifuge the samples at 300 × g for 5 min. The cells form a firm pellet below the saline supernatant. Save the pellet and discard the supernatant by decanting into the sink with running water, taking care not to disturb the cell pellet at the bottom of the tube.
4. Add 500 μl of Chelex beads into the 15 ml test tube containing the cell pellets. Resuspend the cell pellet by either slowly pipetting "in and out" several times or by tapping with your finger.
5. Transfer a 500 μl aliquot of the cell–Chelex slurry into a sterile 1.5 ml Eppendorf tube. Make sure the Eppendorf tube is labeled for identification purposes.

Lysing the Cells and Collecting the DNA

6. Place the capped (closed) Eppendorf tubes in a "float" and place in a boiling water bath for 10 min.
7. After the heat treatment, place the samples on ice for 5 min.

8. Place the Eppendorf tubes containing the lysed cells in a microcentrifuge and spin at the maximum speed for 1 min. The pellet contains the Chelex beads bound to the denatured proteins. The supernatant contains the DNA.
9. Using a 1,000 µl micropipettor with a sterile tip, transfer all of the clear supernatant to a fresh 1.5 ml Eppendorf tube.
10. Label the tube and place on ice until you are ready to proceed to the next step.

Organic Extraction

Organic extraction is a general method used for many situations when stained fabric or clothing is suspected of containing biological material. The stain on the material is cut away from the fabric, soaked in a warm solution (stain extraction buffer) to release the cells from the fabric, incubated with proteinase K, and the DNA isolated using organic solvents. The organic extraction method maintains the integrity of the DNA (i.e., large segments are maintained) while "cleaning" the DNA.

Equipment and Material

1. 2.0 ml screw-capped tube
2. 0.5 ml microcentrifuge tube
3. Adjustable-volume digital micropipets (100–1,000 µl range)
4. Aerosol-resistant pipet tips
5. Stain extraction buffer
6. Proteinase K (10 mg/ml)
7. Phenol:$CHCl_3$:isoamyl alcohol (25:24:1)
8. Chloroform ($CHCl_3$)
9. Disposable gloves
10. Incubator/water bath at 56°C
11. Vortex mixer
12. Microcentrifuge

Procedure

1. Place the cutting from the stained material in a 2.0 ml screw-capped tube.
2. Add 400 µl of stain extraction buffer and 10 µl of proteinase K to each tube. Mix and centrifuge for 2 s. Incubate the tubes containing the cutting at 56°C overnight.
3. Briefly centrifuge the samples.
4. Punch a hole in the bottom of a 0.5 ml microcentrifuge tube. Remove the cutting from the 2.0 ml tube using sterile forceps and place in the 0.5 ml tube. Place the 0.5 ml tube into the 2.0 ml tube from which the cutting was removed.
5. Centrifuge the 2.0 ml tube containing the 0.5 ml inserted tube at maximum speed for 5 min in the microcentrifuge.
6. Remove the 0.5 ml tube and save the cutting.
7. Replace the screw cap on the 2.0 ml tube.

Note: The following steps should be carried out in an exhaust hood.

8. Add 500 µl to each tube, vortex the tube for 20 s, and centrifuge in the microcentrifuge for 2 min.
9. Transfer the top aqueous layer containing the DNA to a new 2.0 ml tube. Do not disturb the interface. Dispose of the phenol:$CHCl_3$:isoamyl alcohol solution in the collection tube in a biohazard waste container.

Steps 10 and 11 are optional unless using Centricon concentration.

10. Add 500 µl of $CHCl_3$ to each tube, vortex, and centrifuge for 2 min.
11. Transfer top aqueous layer to new 2.0 ml tube. Dispose of the $CHCl_3$ solution in the collection tube in a biohazard waste container. The sample is now ready for precipitation or concentration.

"Salting Out"

The salting out procedure is relatively easy to use with liquid samples (known or evidentiary samples) and with cell cultures that might be used as mock evidence samples or as controls. The salting-out DNA isolation procedure involves the preferential hydrolysis and precipitation of cellular proteins. The protein-free genomic DNA is subsequently recovered by either method described in Exercise 7 or 8.

Equipment and Material

1. TE-9 buffer
2. Proteinase K
3. Adjustable-volume digital micropipets (20–200 µl range)
4. Aerosol-resistant pipet tips
5. Cell culture (T-75 flask)
6. 15 ml conical tubes
7. 50 ml conical tubes
8. Disposable gloves
9. 10% sodium dodecyl sulfate (SDS)
10. Saturated NaCl
11. TE buffer
12. 1× Trypsin-EDTA
13. 1× PBS
14. Ice in buckets
15. Inverted microscope
16. Incubator/water bath at 48°C
17. Table top clinical centrifuge

Procedure

1. Decant the growth medium from the cell culture flask. Place the growth medium in a 15 ml conical tube and save for step 5.
2. Wash the cell monolayer twice with 1 ml of 1× PBS (free of calcium and magnesium), decant, and discard.
3. Add 2 ml of 1× trypsin-EDTA to each cell culture flask. Incubate the flask in the palm of your hands for 30–60 s. Under the inverted microscope, observe the "rounding up" of the cells.

Note: In the absence of a microscope, the rounding up of the cells can be assessed by holding the flask up to a light source. As the refractive index changes due to the cells rounding up, the bottom of the flask (which the cells are attached to) will appear cloudy or foggy. The extent of this foggy appearance will depend on the degree of cell rounding and the density of the cell population. It is extremely important not to lyse the cells in the presence of trypsin.

4. To completely dislodge the cells, strike the flask against the palm of your hand. It might be necessary to strike the flask several times against your hand to completely dislodge the cells.
5. To inactivate the trypsin, add 5 ml of the saved medium from step 1.
6. Transfer the cell suspension to a 15 ml conical tube (or suitable centrifuge tube) and centrifuge at 150–200 × g for 3–5 min.

Note: If you plan on concentrating the DNA using the DNeasy Blood and Tissue Kit, go directly to Exercise 7, "Concentration of DNA Using the DNeasy Blood and Tissue Kit" section, step 2.

7. Decant the supernatant and resuspend the cell pellet in 4.5 ml of TE-9 buffer. Add 500 µl of 10% SDS and invert the tube to mix.
8. Add 125 µl of proteinase K to each tube and invert to mix. Incubate the samples at least 30 min at 48°C.
9. Add 1.5 ml of saturated NaCl solution to each tube and shake for 15 s. The lysate should become and remain cloudy.
10. Centrifuge the sample(s) at 500 × g for 10 min to pellet the unwanted proteins.
11. Decant the supernatant containing the DNA into a fresh 15 ml tube. Centrifuge for an additional 10 min.
12. Decant the supernatant containing the DNA into a fresh 50 ml conical tube and place on ice (see Exercise 7 to concentrate the DNA).

Differential Extraction

Differential extraction is the method of choice when biological samples are suspected of containing cells from more than one contributor. Differential extraction is commonly used to isolate the male and female components from a sample containing DNA from a male and a female contributor. Consequently, differential extraction is used to separate sperm cells from nonsperm cells in sexual assault cases. This nonsperm category includes epithelial cells (or skin cells) found in saliva, buccal swabs, vaginal swabs, urine, and feces. The different properties of sperm cells are exploited to separate sperm from these nonsperm or epithelial cells. The separation of the sperm (sometimes referred to as the male fraction) from the epithelial cells (referred to as the female fraction) provides a DNA profile that is easier for the forensic DNA analyst to interpret in a rape case.

Equipment and Material

1. TNE buffer
2. 20% sarkosyl
3. Sterile deionized water
4. Proteinase K (20 mg/ml)
5. Adjustable-volume digital micropipets (20–1,000 µl range)
6. Aerosol-resistant pipet tips
7. 1 M Dithiothreitol (DTT)
8. 2.0 ml screw-capped microcentrifuge tube
9. 0.5 Microcentrifuge tube
10. 2.0 ml screw-capped tubes (conical tubes)
11. Disposable gloves
12. Phenol:$CHCl_3$:isoamyl alcohol (25:24:1)
13. Stain extraction buffer
14. Vortex mixer

15. Microcentrifuge
16. Incubator/water bath at 37 and 56°C

Procedure

Extraction of DNA from Mixtures (or Mixed Stains)

1. The questioned sample containing a stain thought to contain a mixture of sperm and epithelial cells is placed in a 2.0 ml screw-capped Eppendorf/microcentrifuge tube.
2. Mild detergents are then added to remove the stain containing the cells from the material. Add:
   - 400 μl of TNE
   - 25 μl of 20% sarkosyl
   - 75 μl of sterile deionized water
   - 5 μl of 20 mg/ml proteinase K
3. Mix the sample and centrifuge for 2 s. Incubate the sample at 37°C for 2 h.
4. Punch a hole in the bottom of a 0.5 ml microcentrifuge tube. Place stained swab into the 0.5 ml tube and place the 0.5 ml tube into the 2.0 ml tube from which the stained swab was removed. Align the tab on the 0.5 ml tube with the case number label on the 2.0 ml tube.
5. Spin the tube for 5 min.
6. Remove the 0.5 ml tube and place swab in a clean 2.0 ml screw-capped tube. This is fraction 2 (F2).
7. Transfer supernatant to clean 2 ml screw-capped tube. This is fraction 1 (F1). Set aside. The remaining pellet in the tube is the male fraction (M).
8. Add the following components to the pellet labeled (M):
   - 150 μl of TNE
   - 150 μl of H_2O
   - 50 μl of 20% sarkosyl
   - 40 μl of 1 M Dithiothreitol (DTT)
   - 10 μl of 20 mg/ml proteinase K
9. Mix and incubate at 37°C for 2 h.
10. Add 500 μl of stain extraction buffer to the swab (F2).
11. Incubate overnight at 56°C.

Note: The following steps should be carried out in an exhaust hood.

12. Add 400 μl of phenol:$CHCl_3$:isoamyl alcohol to each male fraction and 500 μl of the same to each female fraction.
13. Vortex and centrifuge for 2 min.
14. Transfer the top aqueous layer to a new tube. Dispose of the phenol:$CHCl_3$:isoamyl alcohol solution in the collection tube and dispose of the test tube in the appropriate biohazard waste container.

Steps 15, 16, and 17 are optional unless using Centricon concentration.

15. Add 500 μl of $CHCl_3$ to tube.
16. Vortex and spin for 2 min.
17. Transfer supernatant to new tube or remove bottom layer and discard. Do not disturb the interface. Dispose of the $CHCl_3$ solution in the collection tube. Dispose of the test tube in an appropriate biohazard waste container. The sample is now ready for precipitation or concentration.

DNeasy Blood and Tissue Kit

The DNeasy Blood and Tissue Kit (QIAGEN, Inc., Valencia, California) is designed for the rapid isolation, purification, and concentration of total DNA from animal tissue or cells. The buffer system that is supplied by the manufacturer allows for direct cell lysis, followed by selective binding of the DNA to a silica gel membrane. The lysate is loaded onto the DNeasy mini-column and briefly centrifuged. During centrifugation, the DNA binds to the membrane in the mini-column, while contaminants and enzyme inhibitors (e.g., proteins, divalent cations) pass through the membrane into a collection tube. Following two wash steps, the DNA is eluted in water or a buffer and ready for use (Figure 9.1). The entire procedure and protocol is presented in Exercise 7.

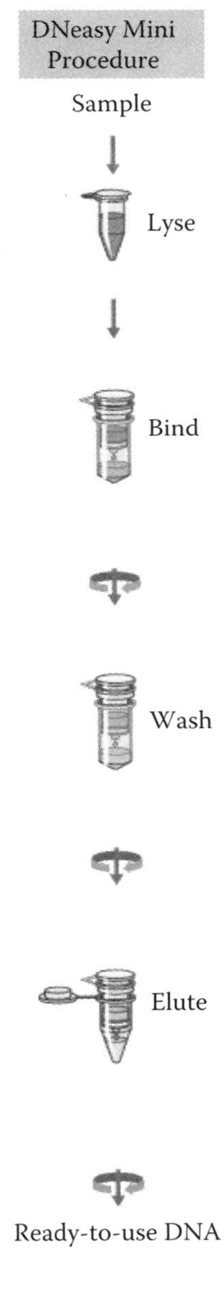

Figure 9.1
DNeasy Blood and Tissue Kit procedure for the isolation, purification, and concentration of DNA. (Courtesy of QIAGEN, Inc.)

Results

Before any analysis proceeds, it is important to determine the success of your extraction and to determine the quality and quantity of DNA present. It is also important to determine if any degradation of the DNA has occurred. The answers to these questions as well as guidelines for interpreting your results are described in Exercise 9.

Samples Extracted					
Analyst:			Lab number:		
Date:					
Item Number/Description	Extraction Method	Purification Method	Item Number/Description	Extraction Method	Purification Method
C = Chelex extraction, S = salting-out extraction, O = organic extraction, D = differential extraction.					
Comments:					

Questions

1. What are the factors that a DNA analyst considers when determining the isolation and extraction procedure to use when analyzing a sample?

2. In your attempt to extract DNA from various samples (both known and evidentiary), several isolation techniques were explored. What technique(s) or method of choice would be used if an evidentiary sample were suspected to contain sperm? Why?

3. In all of the extraction procedures discussed, proteinase K or a detergent (SDS or sarkosyl) was used in the process. What is the purpose of proteinase K? What is the purpose of the detergents?

Chapter 10

Exercise 7
Concentration of Extracted DNA

Introduction

Following the removal and purification of the DNA from the sample, the next step is to concentrate the DNA. Various methods exist to concentrate DNA. Two widely used methods include concentrating extracted DNA by precipitation with ethanol or using a column filtration system (DNeasy Blood and Tissue Kit, QIAGEN, Inc.) to concentrate the DNA. Both techniques are rapid and are quantitative even with nanogram amounts of DNA.

Objective

The objective is to concentrate isolated DNA using two different techniques: ethanol precipitation and column filtration (DNeasy Blood and Tissue Kit).

Equipment and Material

1. Phosphate-buffered saline (PBS)
2. 1× Tris-EDTA (TE) buffer
3. DNeasy Blood and Tissue Kit
4. 1.5 ml Eppendorf (microcentrifuge) tubes
5. Adjustable-volume digital micropipets (2–200 μl range)
6. Aerosol-resistant pipet tips
7. EtOH (70% and 96–100%)
8. Disposable gloves
9. Ice in buckets, –20°C, or –70°C freezer
10. Incubator/water bath at 37, 56, and 70°C
11. Microcentrifuge
12. Tabletop clinical centrifuge

Procedure

The use of disposable gloves and aerosol-resistant pipet tips is highly recommended to prevent cross-contamination. Helpful organizational sheets are provided at the end of the exercise.

Precipitation of DNA Using Ethanol

1. Estimate the volume of the DNA solution and add exactly two volumes of ice-cold absolute ethanol (EtOH). The extracted samples from Exercise 5 contain approximately 500 µl of DNA solution. Add 1 ml of cold absolute EtOH to these samples or tubes (containing the aqueous layer). Mix by hand.
2. Place the tube containing the EtOH/sample on ice for 30 min OR place tube at –70°C for 30 min. Usually 30–60 min at –20°C is sufficient to allow the DNA precipitate to form.
3. Centrifuge the DNA solution containing the EtOH for 15 min. For most purposes, 10 min using a microcentrifuge or at 12,000 × g is sufficient. After centrifugation, decant EtOH.
4. Rinse the DNA pellet with 1 ml of 70% EtOH (room temperature) and centrifuge for 10 min. After centrifugation, decant EtOH.
5. Stand the tube in an inverted position on a layer of absorbent paper until dry (approximately 30 min) or air-dry samples in a secure place.
6. Dissolve the DNA pellet in 36 µl or the desired volume of 1× TE.
7. Resuspend the DNA at 56°C for no more than 2 h. To assist in dissolving the pellet, the sample can be heated to 37°C.
8. Store the sample at 4°C in an Eppendorf or microcentrifuge tube.

Concentration of DNA Using the DNeasy Blood and Tissue Kit

1. Centrifuge the sample for 5 min at 300 × g at room temperature. If human cell lines (e.g., HepG, HeLa, K562) are used, centrifuge approximately 1–5 × 10^6 cells/ml under the same conditions. Cell lines that are archorage dependent will need to be trypsinized prior to harvesting (see Exercise 6, "Salting Out" section).
2. After centrifugation, decant the supernatant and add 200 µl of PBS to the pellet.
3. Add 20 µl of proteinase K and 200 µl of Buffer AL (provided by the manufacturer of the DNeasy Blood and Tissue Kit) to the resuspended cell pellet (see Figure 9.1). Mix thoroughly and incubate at 70°C for 10 min.
4. Add 200 µl of 96–100% EtOH to the sample and mix thoroughly. The sample containing Buffer AL should be mixed thoroughly with the EtOH to ensure a homogenous solution. A white precipitate may form with the addition of EtOH.
5. Place the DNeasy spin column in a 2 ml collection tube (spin columns and collection tubes are provided in the kits by the manufacturer).
6. Place the extracted DNA (or mixture from step 4) into the spin column and centrifuge at greater than 6,000 × g at room temperature for 1 min. The "flow-through," which contains the unwanted cellular material, is discarded along with the collection tube.
7. The DNeasy spin column containing the DNA is placed in a *new* 2.0 ml collection tube (provided in kit).
8. Add 500 µl of Buffer AW1 to the spin column and centrifuge the column/tube at 6,000 × g at room temperature for 1 min.
9. Following centrifugation, the flow-through again is discarded, along with the collection tube and the spin column placed in a new 2.0 ml collection tube.
10. Pipet 500 µl of Buffer AW2 into the spin column and the column/tube centrifuged at full speed for 3 min. This centrifugation step ensures that no residual EtOH is carried over during the following elution.
11. Following centrifugation, the flow-through is discarded and the spin column placed in a 1.5 or 2 ml microcentrifuge tube. Pipet 200 µl of Buffer AE onto the column and incubate at room temperature for 1 min.
12. The spin column/microcentrifuge tube is then centrifuged at 6,000 × g for 1 min to elute the DNA.

Results

Before any analysis proceeds, it is important to determine the success of this concentration procedure. It is important to determine the quality and quantity of DNA present. It is also important to determine if any degradation of the DNA has occurred. The answers to these questions as well as guidelines for interpreting your results are described in Exercise 9.

Precipitation of DNA Using Ethanol					
Analyst:			Lab number:		
Date:					
Item Number/Description	Extraction Method	Purification Method	Item Number/Description	Extraction Method	Purification Method
Comments:					

QIAGEN Extraction/Purification Method: Manual Spin Columns	
Analyst:	Date:
System:	Gel number:
Reagent	**Lot/Source**
1× PBS	
Protease	
Buffer AL	/QIAGEN
95% EtOH	
Buffer AW1	/QIAGEN
Buffer AW2	/QIAGEN
Buffer AE	/QIAGEN
Lysate transfer witness:	
Elution tube transfer witness:	

Questions

1. What is the purpose of concentrating the extracted DNA?

2. Two different concentration techniques were discussed: ethanol precipitation and column filtration. What are the advantages and disadvantages when using these techniques to concentrate DNA?

Chapter 11

Exercise 8

Microcon Concentration and Purification of Extracted DNA

Introduction

After DNA extraction, centrifugal filter devices, such as the Microcon purification procedure, can serve as powerful tools in DNA concentration and desalting procedures. Ultrafiltration (UF) is a pressure-driven, convective process that uses semipermeable membranes to separate DNA by molecular size and shape. Ultrafiltration is highly efficient, allowing for concentration and purification at the same time. Unlike the use of chemical precipitation methodologies (i.e., as in Exercises 6 and 7 using phenol/chloroform or ethanol), there is no phase change or possible degradation of the DNA with UF. Ultrafiltration routinely concentrates DNA, without the use of co-precipitants, in a short time period with 99% recovery of the starting material. Centrifugal concentrator devices are ideal for separating high and low molecular weight DNA molecules.

The Microcon purification procedure is often used when the biological sample that was extracted was deposited on a substrate (i.e., denim or velvet) known to inhibit DNA amplification or the polymerase chain reaction (see Exercise 10) due to the substrate releasing excessive amounts of dye during the extraction process. In this exercise, the DNA sample is concentrated, then diluted to the original volume with the desired buffer, and concentrated again, thus "washing out" inhibitors or the original solvent.

Objective

The purpose of this exercise is to concentrate and remove unwanted cellular or foreign components from the isolated DNA using ultrafiltration and the Microcon purification procedure.

Equipment and Material

1. Microcon 100 concentrator assembly (Millipore Corporation)
2. Adjustable-volume digital micropipets (2–200 µl range)
3. Aerosol-resistant pipet tips
4. 1× TRIS-EDTA (TE) buffer
5. 1.0 or 1.5 ml microcentrifuge tubes
6. Sterile H_2O

7. Transfer pipets
8. Disposable gloves
9. Phenol:$CHCl_3$:isoamyl alcohol (25:24:1)
10. Bromophenol blue tracking dye (loading buffer)
11. Microcentrifuge

Procedure

The use of disposable gloves and aerosol-resistant pipet tips is highly recommended to prevent cross-contamination. A helpful organizational sheet is provided at the end of the exercise.

Concentrating the DNA Using a Microcon Concentrator

1. Use a Microcon concentrator that is adequate for your DNA size (use a Microcon 50 concentrator for minute biological stains/materials). Add 500 µl of prewarmed (room temperature) phenol:$CHCl_3$:isoamyl alcohol to each tube containing the DNA.
2. Cap the tube and mix thoroughly by hand for 2–3 s or until solution has a milky appearance.
3. Centrifuge the tube(s) for 3 min in a microcentrifuge at greater than 6,000 × g to separate the two phases.
4. Insert a labeled Microcon 100 concentrator into a labeled collection vial. Add 500 µl of sterile H_2O to the concentrator. Using a transfer pipet, transfer the aqueous or top phase containing the DNA (from step 3 above) to the Microcon concentrator. Place the cap from the collection vial on the concentrator.
5. Centrifuge the Microcon assembly in a microcentrifuge for 10–30 min at approximately 5,000 rpm (× g or relative centrifugal force (RCF) determined by rotor used) until the volume is reduced.
6. After centrifugation, remove the concentrator "unit" from the Microcon assembly and discard the fluid from the filtrate cup. Return the concentrator to the top of the filtrate cup.
7. Remove the cap from the concentrator and add 200 µl of sterile H_2O. Replace the cap and centrifuge at 5,000 rpm (× g or RCF determined by rotor used) for 10–30 min until the volume is reduced.
8. Remove the cap from the concentrator and add 30 µl of 1× TE buffer.
9. Remove the concentrator from the filtrate cup and carefully invert the concentrator onto the retentate cup. Discard the filtrate cup.
10. Centrifuge the Microcon assembly (the retentate cup end first) at 5,000 rpm (× g or RCF determined by rotor used) for 5 min.

Note: Dissolved components that pass through the membrane are known as permeate. The components that do not pass through are known as retentate.

OPTIONAL

11. Remove 4 µl of sample and place in a separate 1.0 ml tube. Add 2 µl of loading buffer to the sample (see Exercise 9).
12. Run the sample on a test agarose gel for quantitation (see Exercise 9).

Results

Before any analysis proceeds, it is important to determine the success of your extraction, purification, and concentration of the DNA. It is important to determine the quality and quantity of DNA present. It is also important to determine if any degradation of the DNA has occurred. The answers to these questions as well as guidelines for interpreting your results are described in Exercise 9.

Microcon Concentration and Purification of Extracted DNA					
Analyst:			Lab number:		
Date:					
Item Number/Description	Extraction Method	Purification Method	Item Number/Description	Extraction Method	Purification Method
Comments:					

Questions

1. What are the advantages of the Microcon concentrator assembly over the previous extraction processes described?

2. What are the advantages over the two techniques (ethanol precipitation and column filtration) described previously to concentrate extracted DNA?

Chapter 12

Exercise 9

Assessing the Quality and Quantity of Isolated DNA

Introduction

To determine the quality and quantity of extracted or isolated DNA recovered from a sample (known or evidentiary), preliminary tests are conducted. Two tests are often used to assess the amount (how much DNA is present) or quality (how much, if any, degradation has occurred) of the DNA.

In the first test, a miniature agarose gel or yield gel is used to estimate both the quality and quantity of DNA recovered from each sample. A yield gel is prepared and a small portion of each DNA sample is loaded into separate wells of the gel. The DNA is analyzed by agarose gel electrophoresis, stained to visualize the DNA by UV illumination, and photographed (or the images generated using computer software). Documentation for each gel is maintained and indicates the DNA samples that have been included on a particular yield gel, along with the appropriate controls (i.e., visual marker, HindIII-digested lambda DNA; human K562 DNA control (Lozzio and Lozzio, 1979) or intact lambda DNA for calibration or quantity determination). Large, intact, and undegraded DNA will appear as a compact band near the origin of the gel, similar to the standards in the adjacent lanes. Degraded DNA will form a smear and will migrate further through the gel, depending on the various sizes of the DNA fragments (Figure 12.1, lanes 7–9). Degraded DNA will also be observed following hybridization and autoradiography/chemiluminography (Figure 12.2, lane 3 and Figure 12.3, lane 3). Extremely degraded DNA may not be visible since these smaller fragments will migrate toward the end of the gel. The quantity of the DNA in question can be compared to DNA standards of known quantity that have been run in adjacent lanes.

The second method consists of the slot or dot blot technique that is used to determine only the quantity of the DNA recovered from a sample. A small portion of the sample DNA in question is applied to a membrane along with a set of standard samples of known quantity. After the samples have been fixed to the membrane, a human DNA probe is added and allowed to hybridize to the fixed DNA. The DNA probe used is this instance is tagged or labeled (e.g., enzyme-linked or fluorescent dyes) for easy detection of the DNA (Figure 12.4). The slot or dot blot technique does not provide any information on the quality or the level of degradation of the DNA.

Currently, many forensic laboratories are using the AluQuant™ System (Promega Corporation, Madison, Wisconsin) in conjunction with the BioMek® 2000 Automation Workstation (Beckman Coulter, Fullerton, California) to estimate the concentration of isolated or extracted human DNA. Briefly, the AluQuant System measures DNA through the use of DNA probes that bind to highly repetitive sequences (an Alu element is a short stretch of DNA originally characterized by the action of Alu, a restriction endonuclease) previously identified in human chromosomes, without amplification by polymerase chain reaction

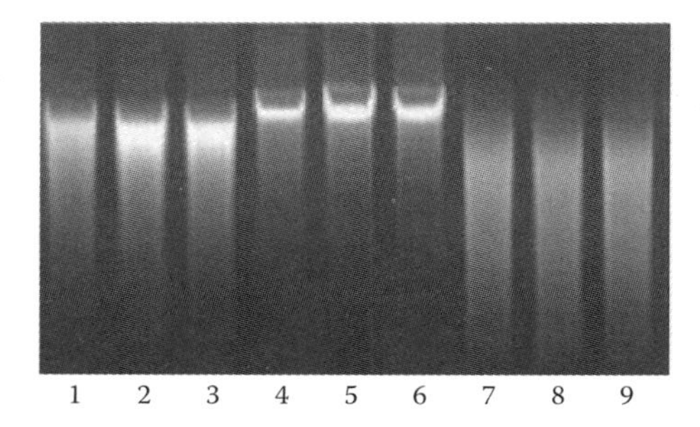

Figure 12.1
Agarose gel electrophoresis demonstrating degraded DNA. Human buffy coat samples were stored for 27 months at room temperature (lanes 1–3) and for 11 years at 45°C (lanes 7–9). Human buffy coat samples were also preserved in DNAstable Blood (Biomatrica, Inc.) for 11 years at room temperature (lanes 4–6). After the buffy coat samples were rehydrated, the genomic DNA was extracted, purified, and analyzed by agarose gel electrophoresis. Lanes 4–6 show undegraded DNA that appears as a compact single band near the origin or top of the gel, whereas degraded DNA (lanes 1–3 and 7–9) appears as a smear. (Courtesy of Biomatrica, Inc., San Diego, CA.)

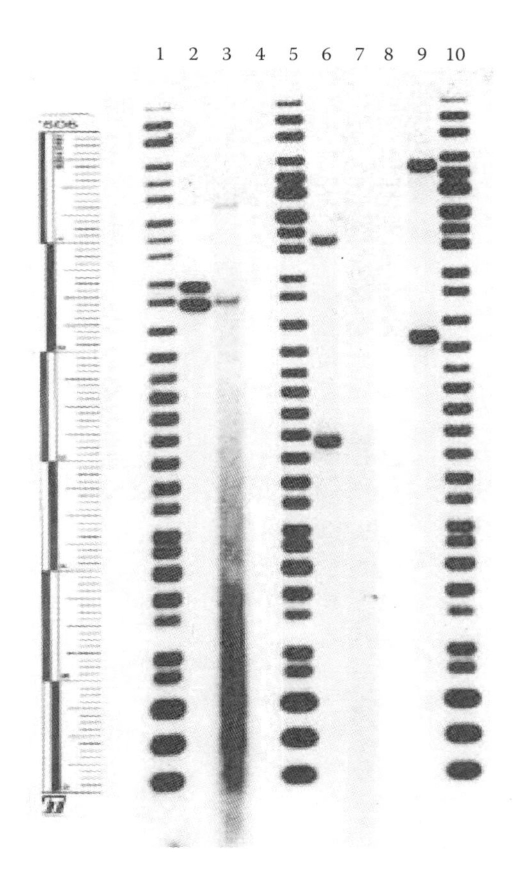

Figure 12.2
A restriction fragment length polymorphism (RFLP) lumigraph demonstrating DNA degradation. Lanes 1, 5, and 10 contain molecular ladders. Lanes 2 and 9 serve as positive controls and contain a known reference sample. Lane 3 contains degraded DNA from the nonsperm fraction from the victim. Lane 6 contains the suspect's reference sample. Lanes 4 and 8 are blank. Lane 7 contains the sperm fraction from the sexual assault sample. The suspect is a heterozygote at this locus since two DNA fragments were observed. However, only one faint DNA fragment (the lower molecular weight fragment) was observed at this locus from the sexual assault sample, which would lead the DNA analyst to possibly render an inconclusive result.

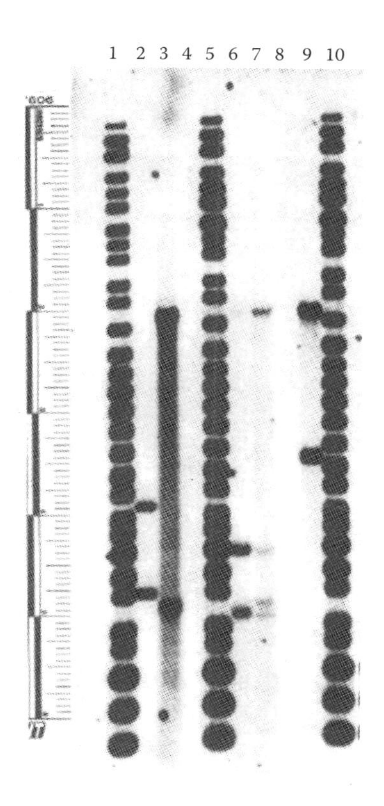

Figure 12.3
A restriction fragment length polymorphism (RFLP) lumigraph demonstrating DNA degradation with two DNA fragments observed. Lanes 1, 5, and 10 contain molecular ladders. Lanes 2 and 9 serve as positive controls and contain a known reference sample. Lane 3 contains degraded DNA from the nonsperm fraction from the victim. However, two DNA fragments are observed. Lane 6 contains the suspect's reference sample. Lanes 4 and 8 are blank. Lane 7 contains the sperm fraction from the sexual assault sample. The suspect's DNA profile is observed as well as the victim's profile in this sexual assault sample. It should be noted that the "carryover" is evident from the nonsperm fraction into the sperm fraction. Also evident in the sperm fraction are DNA fragments that align with the suspect's two-band profile.

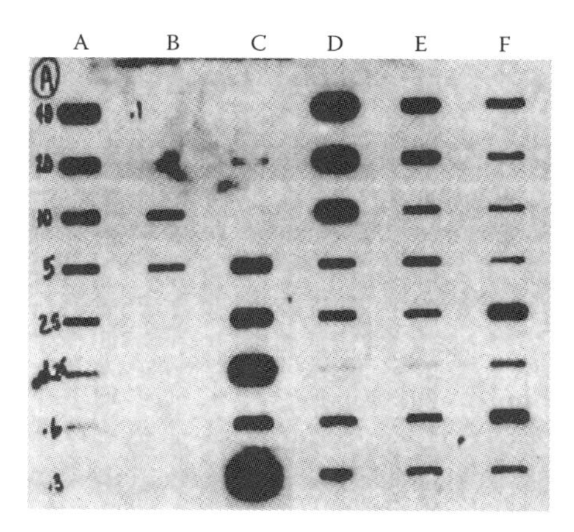

Figure 12.4
A slot blot used to determine the quantity of human DNA recovered from a sample. The first column or lane (A) contains the quantitation standards in decreasing order from top to bottom (40, 20, 10, 5, 2.5, 1.25, 0.6, and 0.3 ng DNA). The remaining lanes contain "responses" from the evidence (lanes B and C) and reference samples (lanes D and E), as well as from the positive (lane F) and negative (bottom of lane B) controls. The amount of DNA contained in a sample (evidentiary or reference) is obtained by comparing the response to the standards.

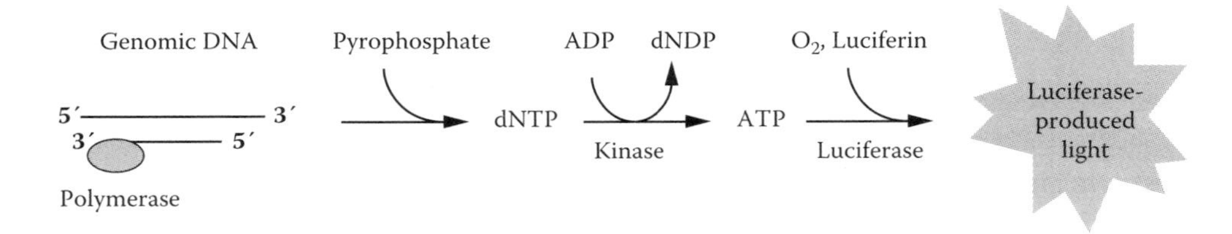

Figure 12.5
The chemical and enzymatic pathway of the AluQuant™ Human DNA Quantification System. The amount of light generated (i.e., luciferase-produced light) is proportional to the amount of ATP present and consequently the amount of human DNA present in the sample. (Courtesy of Promega Corporation.)

(PCR). Immobilization of the DNA target on a substrate (e.g., by Southern blotting) is unnecessary since the hybridization of the DNA probes and target is performed in solution. Following hybridization, a series of enzymatic reactions occurs that produces a luminescent signal that is proportional to the quantity of human DNA present in the sample tested (Figure 12.5). The amount of DNA present in the purified sample can then be determined by comparing the net luminescent signal from the unknown to a standard curve established with known amounts of human DNA. Accuracy of measurement lies in the ranges of 20 pg/μl to 4 ng/μl in the DNA sample being analyzed.

Objective

To determine the quality and quantity of DNA isolated from either a known sample, an evidentiary sample, or a human cell line (e.g., HepG, HeLa, K562) using agarose gel electrophoresis, a simple and rapid technique is utilized in many forensic laboratories.

Equipment and Material

1. Agarose (DNA typing grade)
2. Ethidium bromide (1 μg/ml)
3. Ice in buckets
4. 1× Tris-Borate-EDTA (TBE) buffer
5. K562 DNA
6. 125 ml Erlenmeyer flask
7. Adjustable-volume digital micropipets (2–200 μl range)
8. 1.0 or 1.5 ml microcentrifuge tubes
9. Aerosol-resistant pipet tips
10. Bromophenol blue tracking dye
11. Microwave or hot plate
12. Disposable gloves
13. Power pack/power supply
14. Incubator/water bath at 56°C
15. Electrophoresis systems (gel tray/combs)
16. UV transilluminator
17. Polaroid camera with film or computer-assisted image capabilities

Procedure

The use of disposable gloves and aerosol-resistant pipet tips is highly recommended to prevent cross-contamination. Helpful organizational sheets are provided at the end of the exercise to record your data and observations. An overview of the gel casting process and setting up the electrophoretic run is illustrated in Figures 12.6 and 12.7.

1. Preparation of 0.8% agarose test gel.
    a. Add 0.8 g of DNA typing grade agarose in 100 ml of TBE buffer containing ethidium bromide (EB) at a ratio of 10 μl of EB/100 ml of TBE buffer.

Note: Some laboratories will stain the agarose gel with ethidium bromide after electrophoresis instead of adding ethidium bromide to the TBE containing the agarose or the reservoir buffer.

    b. Heat the agarose solution in a microwave on high for about 40 s, swirling the flask by hand every 10–15 s, or briefly bring to a boil to dissolve the agarose using a hot plate.
    c. Cool the liquid agarose to about 56°C. While the agarose cools, prepare the gel tray according to the manufacturer's guidelines (e.g., place the gel dams at each end of the tray).
    d. Pour the agarose into the gel tray. Use either one or two 14-well combs.

Note: The number of wells (or teeth) in a comb will vary according to the manufacturer.

    e. Let the liquid agarose stand or cool in the gel tray for 10 min to solidify.
    f. Remove gel dams and place agarose gel tray into the electrophoresis chamber or gel tank containing approximately 175 ml of TBE/EB buffer.
    g. Remove the comb(s).

Note: If the gel has solidified, but you are not ready to load the gel, add the TBE buffer anyway so that the gel will not dry out.

2. Remove 4 μl of extracted or resolubilized DNA and combine with 2 μl tracking dye. This may be done in a microcentrifuge tube or in a microtiter plate.
3. The DNA sample, mixed with loading buffer (tracking dye), is pipetted into the well with the gel submerged. The final sample volume should contain 10% tracking dye, and the total volume should not exceed 20 μl. Be careful not to push the pipet tip through the bottom of the well in the gel when loading the sample.
4. Include on your gel the following K562 DNA standards:

    500 ng/4 μl
    250 ng/4 μl
    125 ng/4 μl
    63 ng/4 μl
    31 ng/4 μl
    15 ng/4 μl

Note: Intact lambda DNA can be used in quantities ranging from 10 to 300 ng.

5. Set the voltage (100 V) and "run" the samples until the bromophenol blue tracking dye has moved 1–2 cm from the origin (i.e., well) or until the dye front is approximately 2 cm from the end of the gel. This should take less than 20 min.
6. Remove the gel from the electrophoresis tank. Stain the gel in syber green or ethidium bromide (if not added previously to agarose gel or buffer reservoir).

7. Examine the gel on an ultraviolet (UV) transilluminator. Take a photograph of your gel. Intact DNA will move as a band not far from the origin. A smear from the origin to or past the dye front indicates that the DNA has been fragmented and may not be suitable for further use.

Avoid excessive exposure to the UV light. Always wear a full face shield when working with the transilluminator.

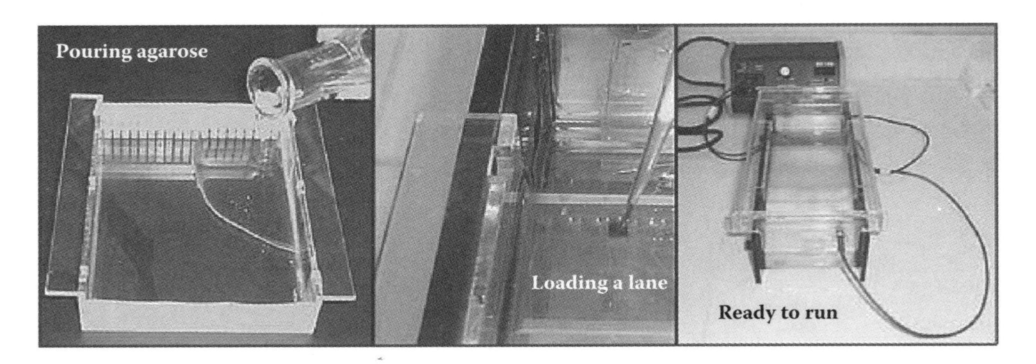

Figure 12.6
Setting up and loading the agarose gel. (a) While the liquid agarose is cooling, the gel tray is prepared by placing the gel dams at each end of the tray. The comb is placed at one end of the gel tray and the agarose poured into the gel tray. (b) After the agarose has solidified, the gel comb and gel dams are removed, the tray placed in the electrophoresis chamber containing buffer, and the DNA samples, containing the tracking dye, are loaded into each well. (c) Voltage is applied and the DNA molecules are separated in an electrical field. (Courtesy of MIT OpenCourseWare.)

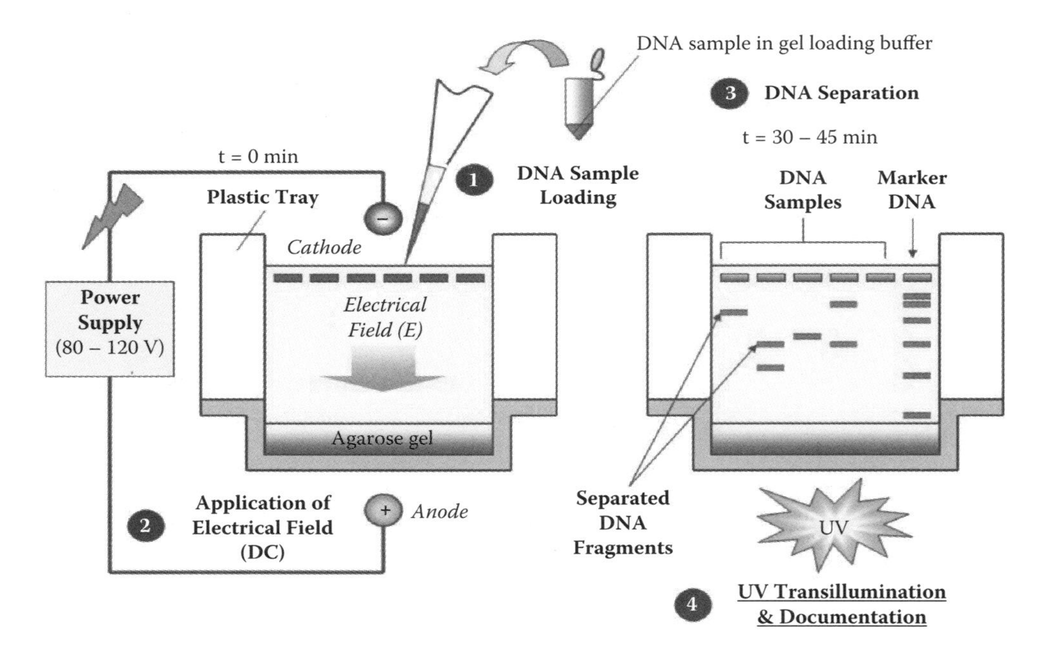

Figure 12.7
An overview of the gel electrophoretic process. (1) The DNA samples are loaded into the wells of the gel. (2) The voltage is set and the samples are run until the tracking dye or dye front is approximately 2 cm from the end of the gel. (3) The DNA molecules are separated and (4) visualized by ethidium bromide (or alternate dye, e.g., SYBR green) staining and UV illumination. (From http://classroom.sdmesa.edu/eschmid/Lab8-B10/210.htm.)

Results

Yield Gel		
Analyst:	Lab number:	
Date:		
Well No.	Sample	DNA (ng)
1	Visual marker	
2	Standard	500
3	Standard	250
4	Standard	125
5	Standard	63
6	Standard	31
7	Standard	15
8		
9		
10		
11		
12		
13		
14		
15		
16		
17		
18		
19		
20		

Reagents	Lot No.	Source
Agarose		
1× TBE (gel buffer)		
1× TBE (tank buffer)		
Loading buffer		
Ethidium bromide		
Visual marker		
500 ng of standard		
250 ng of standard		
125 ng of standard		
63 ng of standard		
31 ng of standard		
15 ng of standard		

Gel Electrophoresis		
Time on:	Voltage:	mAMPs
Time off:	Voltage:	mAMPs

Reporting Form

Tape Your Gel Photo Here	Lane Contents
	Lane 1
	Lane 2
	Lane 3
	Lane 4
	Lane 5
	Lane 6
	Lane 7
	Lane 8
	Lane 9
	Lane 10
	Lane 11
	Lane 12
	Lane 13
	Lane 14
	Lane 15
	Lane 16
	Lane 17
	Lane 18
Gel prepared by:	Lane 19
Date:	Lane 20

Yield Gel		
Analyst:	Lab number:	
Date:		
Well No.	Sample	DNA (ng)
1		
2		
3		
4		
5		
6		
7		
8		
9		
10		
11		
12		
13		
14		
15		
16		
17		
18		
19		
20		

Interpreting Test Results

1. While under UV illumination take a photograph of the gel and attach it to the reporting form.
2. From the photograph, estimate the quality of DNA in the test specimens by comparison to the uncut K562 DNA standards.
   - Intact DNA: Large, intact, and undegraded DNA will appear as a compact band near the origin of the gel, similar to the standards in the adjacent lanes (see Figure 12.1, lanes 4–6).
   - Degraded DNA: Degraded DNA will form a smear along the lane and will migrate further through the gel, depending on the various sizes of the DNA fragments (Figure 12.1, lanes 7–9). Degraded DNA will also be observed following hybridization and chemiluminography (see Figure 12.2, lane 3 and Figure 12.3, lane 3). Extremely degraded DNA may not be visible since these smaller fragments will migrate toward the end of the gel.

- Record the quality of DNA (i.e., intact versus degraded) in the results section using the yield gel form in the column labeled "Sample."

3. From the photograph, estimate the quantity of DNA in test specimens. The quantity of the DNA in question can be compared to DNA standards of known quantity that have been run in adjacent lanes.
   - A single estimate should be made (not a range), and the quantity per 4 and 32 μl of total DNA remaining in the sample should be recorded on the worksheet. The estimation is multiplied by 8 to obtain the total quantity of DNA in the remaining 32 μl of sample. Record the estimated quantity of DNA in the results section using the yield gel form.

Questions

1. Once the DNA has been isolated and before any analysis can proceed, it is imperative to determine the quality and quantity of DNA present. Why is it important to determine the quantity and quality of DNA in a sample?

2. How would a DNA analyst determine, in one experiment, the quantity and quality of DNA from a given sample?

3. If the evidentiary sample to be analyzed was degraded, what are the many forms of DNA that would be expected following gel electrophoresis?

Chapter 13

The Polymerase Chain Reaction (PCR)

Introduction

If the evidentiary sample contains an insufficient quantity of DNA, or if the DNA is degraded, a polymerase chain reaction (PCR)-based test may be used to obtain a DNA profile. The PCR-based tests generally provide rapid results that can serve as an alternative or as a complement to other DNA testing. The process involves the isolation of DNA from a biological specimen (e.g., blood, semen, saliva, fingernail clippings), followed by an assessment of DNA quality and quantity. Next, the PCR amplification technique is used to produce millions of copies of a specific portion of a targeted DNA segment that contains polymorphic DNA selected for forensic and parentage evaluations (Figure 13.1). The PCR amplification procedure is similar to a molecular photocopying machine. The amplified PCR products are then separated and identified by either gel-based or capillary electrophoresis where chemical staining (e.g., ethidium bromide or Coomassie blue) or chemiluminescent labels are used, respectively. The resulting DNA profiles are routinely interpreted by direct comparison to allele standards (i.e., allelic ladders for each locus) and known reference samples. Probability calculations are determined based upon classical population genetic principles.

Early forensic detection systems relied on the quality and quantity of the DNA sample to be analyzed. The large amount (a dime-sized stain) of isolated DNA, whether from an evidentiary or a known reference sample, had to be relatively fresh or undegraded, essentially unadulterated for these detection systems to yield a sufficient profile. For samples considered too minuscule (i.e., low concentration) or determined to be degraded, the PCR or amplification process is now performed.

DNA typing using PCR and gel electrophoresis eliminated the need for critically sensitive DNA probes previously used with the restriction fragment length polymorphism (RFLP) procedure. Analysis time ranges between 24 and 48 h. Computer-assisted image analysis of test results is helpful but not always necessary since the resulting genetic profiles are routinely interpreted by visual or direct comparison to allele standards at specific loci. Population frequencies are conservative estimates based upon classical population genetic principles.

The PCR methodology also has demonstrated consistency of results from tissue to tissue, body fluid to body fluid, within an individual. Therefore, the detection of an allele(s) in an unknown biological stain allows for comparison to known reference samples regarded as a possible source or contributor to such biological materials. If the collection of genetic information (DNA profile) associated with an unknown stain is consistent with the results generated from a known reference sample, then the possibility that a common genetic source exists for both sets of samples cannot be eliminated. The demonstration of independent Mendelian inheritance, as mentioned earlier, allows for a conservative estimate of the frequency of such a DNA profile occurring among unrelated individuals randomly selected in various racial groups.

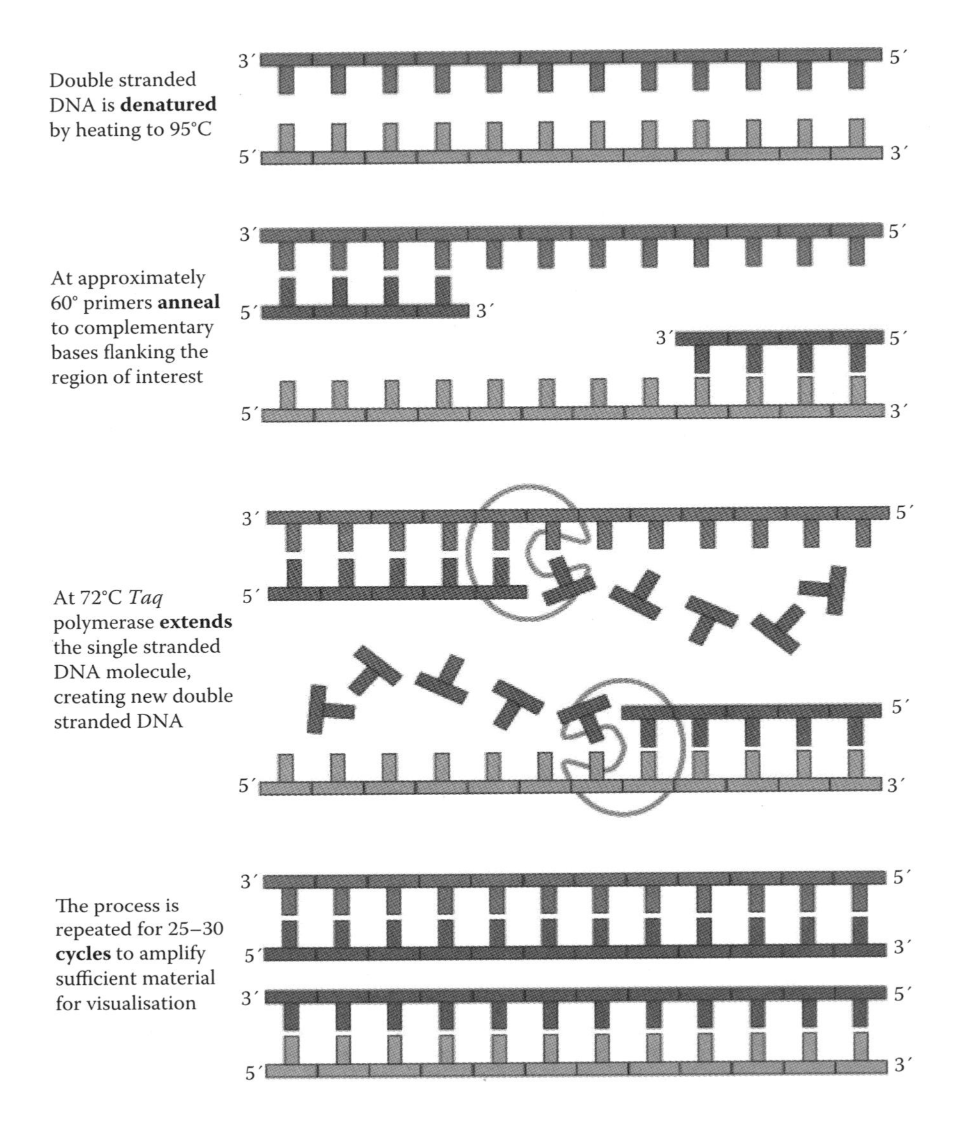

Double stranded DNA is **denatured** by heating to 95°C
At approximately 60° primers **anneal** to complementary bases flanking the region of interest
At 72°C *Taq* polymerase **extends** the single stranded DNA molecule, creating new double stranded DNA
The process is repeated for 25–30 **cycles** to amplify sufficient material for visualisation
3′
5′

Figure 13.1
An overview of the polymerase chain reaction (PCR). A defined region of DNA is copied using the PCR amplification technique to produce millions of copies of the specified region on the DNA strand. (Courtesy of the University of Leicester, East Midlands Forensic Pathology Unit, Leicester, UK. http://www2.le.ac.uk/department/emfpu/genetics/explained/mitochondrial.)

Objective

Exercises 10–12 will introduce the student to the PCR amplification procedure and short tandem repeat (STR) analysis, a PCR-based test.

Laboratory Setup

Due to the sensitivity of PCR-based tests, certain precautions are necessary to avoid contamination of samples with other sources of DNA. To minimize the potential for laboratory-induced DNA contamination, several aspects of the PCR process should be considered: (1) DNA extraction, (2) PCR setup, and (3) amplified DNA analysis. Each aspect of the PCR process should be separated by time and space. The following list addresses special precautions that must be taken to minimize contamination in the laboratory.

1. The work area for DNA extractions and nonamplified DNA should include dedicated equipment and supplies.
2. DNA extractions and PCR setup should be conducted within self-contained hoods. If hoods are unavailable, use an area of a benchtop that is dedicated for this use only.
3. Use disposable gloves at all times and change frequently. Prior to leaving the laboratory area, always remove the gloves and wash your hands.
4. DNA extraction of questioned samples (i.e., evidentiary samples) should be performed separately from the extraction of known samples. This will minimize the potential for cross-contamination between samples.
5. Every sample to be analyzed should be properly labeled and recorded. Evidentiary and known reference samples to be analyzed in a forensic laboratory are given a unique identification number that is used throughout the entire analysis.
6. To minimize sample-to-sample contamination, extract samples containing high levels of DNA (e.g., whole blood) separately from samples containing a low level of DNA (e.g., small bloodstains, stamps, and envelopes).
7. Always use sterile solutions or reagents and, whenever possible, sterile disposable supplies (i.e., pipet tips, microcentrifuge tubes).
8. Always change pipet tips between handling each sample even when dispensing reagents.
9. Sterilize reagents and store as small aliquots to minimize the number of times a given tube of reagent is opened. It is recommended that the small aliquots be retained until typing of the set of samples for which the aliquots were used is complete. Then dispose of the tube containing the reagent.
10. Include reagent blank controls with each set of DNA extractions.
11. Before and after setting up the DNA extractions, clean all work surfaces thoroughly with a 10% solution of bleach. In addition, the use of disposable bench paper will prevent the accumulation of human DNA on permanent work surfaces.

Chapter 14

Exercise 10
Short Tandem Repeat (STR) Analysis

Introduction

A DNA segment that appears more than once on the same chromosome is known as a repeat. Human genomes contain 5–10% of such repetitive sequences that occur in tandem or adjacent to each other. These repetitive sequences vary in size and length and show sufficient variability among individuals in a population. Regions of DNA that contain these short repeated segments are referred to as short tandem repeats (STRs) and are important markers for human identity testing in the forensic community.

There are literally thousands of STR markers scattered throughout the human genome, and they occur, on average, in 1 in every 10,000 nucleotides. The DNA sequence repeated in an STR motif is usually from 2 to 6 base pairs (bp), with four bases being the preferred size for forensic systems (Edwards et al., 1991, 1992; Warne et al., 1991). An example of a 4 bp or a tetranucleotide repeat is shown below, where the TCTA motif is repeated four times.

...ATGTGA[TCTA]TCTATCTATCTATTGG...

Polymerase chain reaction (PCR)-based STR systems offer many advantages over earlier DNA typing techniques (e.g., restriction fragment length polymorphisms [RFLPs]). STR systems provide a rapid and sensitive method to evaluate small amounts (1 ng) of human DNA. This small amount of DNA needed for STR systems is 50 times less than what is normally required for RFLP analysis. Also, the repeating sequences in an STR are relatively short, with the entire STR strand or allele generally less than 400 bp in length. This short length renders STR systems amenable to the analysis of samples suspected of being degraded. STR analysis often allows the DNA analyst to recover a complete DNA profile even from an evidentiary sample that was exposed to unfavorable conditions (e.g., body or stains subject to extreme decomposition). This is in sharp contrast to RFLP systems that required a large sample size for analysis and full-length fragments, which often consisted of thousands of bases, to generate a complete DNA profile.

STRs and corresponding loci are easily amplified by PCR. Further, PCR amplification of many different STR loci is commonly performed simultaneously in the same tube. The simultaneous amplification of two or more loci is commonly known as multiplexing or multiplex PCR. For a multiplexing reaction to be successful, the system must be designed to ensure that the size of the amplified products does not overlap, thereby allowing each STR allele for a specific locus to be clearly visualized on a gel or by capillary electrophoresis. This requirement design of overlapping fragments became less important with the development of multiple color detection systems.

Different detection methods are available to visualize the STR products. The STR loci and corresponding alleles may be separated by gel electrophoresis and detected using ethidium bromide and silver staining

TABLE 14.1
Information on the 13 Core Short Tandem Repeat Loci Listed in CODIS

STR Locus	Chromosome Number	Sequence
FGA	4	CTTT
vWA	12	[TCTG][TCTA]
D3S1358	3	[TCTG][TCTA]
D21S11	21	[TCTA][TCTG]
D8S1179	8	TATC
D7S820	7	GATA
D13S317	13	TATC
D5S818	5	AGAT
D16S539	16	GATA
CSF1PO	5	AGAT
TPOX	2	AATG
THO1	11	TCAT
D18S51	18	AGAA

or exotic dyes (e.g., SYBR green). Several STR systems have been developed where fluorescent dyes or labels are used to detect the STR alleles either during (i.e., capillary electrophoresis) or after (i.e., gel electrophoresis) separation. The resulting STR profiles are routinely interpreted by direct comparison to DNA standards, allelic ladders (an artificial mixture of common alleles present in the human population for a particular STR marker or locus), and reference standards (known DNA profiles from the victim and suspect). Probability calculations are determined based upon classical population genetic principles.

For STR markers to be effective across various jurisdictions, a common set of standardized markers is used. Currently, the forensic scientific community in the United States has established a set of 13 core STR loci that, in turn, can be entered into a national database known as the Combined DNA Index System (CODIS), a collection of DNA profiles from known offenders. A summary of the 13 CODIS loci is contained in Table 14.1.

Objective

In this exercise, you will extract DNA from cells in your mouth (buccal swabs) or from a human cell line, incubate the isolated DNA with appropriate PCR reagents, and amplify the alleles at multiple STR loci using the GenePrint STR Systems (Promega Corporation, Madison, Wisconsin). Following amplification, the amplified STR products will be separated and identified using agarose gel electrophoresis.

Equipment and Material

1. 0.5, 1.0, and 1.5 ml Eppendorf/microcentrifuge tubes
2. 30 or 50 ml conical tubes
3. Mineral oil (optional)
4. 15 ml polypropylene test tube
5. Double-distilled sterile water
6. Sterile cotton swabs
7. Adjustable-volume digital micropipets (2–200 μl range)

8. Aerosol-resistant pipet tips
9. GenePrint STR Systems CSF1PO, TPOX, TH01 (Promega Corporation, Madison, Wisconsin)
10. *Taq* DNA polymerase (not supplied in kit)
11. Disposable gloves
12. Ice in buckets
13. Genomic DNA from human cell lines* (10 ng/μl)
    a. HEP G2—hepatocellular carcinoma (liver), male
    b. HTB 180 NCI-H345—small cell carcinoma, lung
    c. CCL 86—Raji Burkitt lymphoma
    d. HTB 184 NCI-H510A—small cell carcinoma, extra pulmonary origin
    e. CRL 1905 H—normal skin cell line
    f. HeLa—epithelial carcinoma cell line
    g. K562—erythromyeloblastoid leukemia cell line, chronic myeloid leukemia
    h. GM9947A—human lymphoid cell line
14. Molecular weight markers (526–22,621 bp)
15. Ethidium bromide/Coomassie blue
16. Bromophenol blue tracking dye
17. 125 ml Erlenmeyer flask
18. 1× TBE buffer
19. Agarose (DNA typing grade)
20. DNA thermal cycler
21. Electrophoresis systems (gel tray/combs)
22. Power pack/supply
23. Microwave or hot plate
24. Incubator/water bath at 56°C
25. Microcentrifuge
26. Tabletop clinical centrifuge
27. Polaroid camera with film or computer-assisted imaging capabilities

Procedure

To prevent cross-contamination, the use of disposable gloves and aerosol-resistant pipet tips is highly recommended. Helpful organizational sheets are provided at the end of the exercise.

1. Refer to Exercise 6 for the methods and steps used in the collection and concentration of cells and for methods outlining cell lysis and the collection of DNA.

* Examples of human cell lines that can be used to demonstrate an STR profile. The STR profiles are available through the American Type Tissue Collection (ATCC) at http://www.atcc.org/.

Setting Up the PCR Amplification

1. Thaw the PCR reagents (STR 10× buffer and STR 10× primer pairs) and keep on ice. These reagents will be combined to form the PCR master mix for the multiplex reactions.
2. Determine the number of reactions to be set up. Positive and negative controls should also be included when determining the number of reactions.
3. For each reaction, label one sterile 0.5 ml microcentrifuge tube and place into a rack.
4. To determine the final volume of the master mix needed for all of the reactions, calculate the required amount of each component of the PCR master mix (see Table 14.2). Multiply the volume (μl) per sample by the total number of reactions (from step 2) to obtain the final volume (μl). To compensate for pipetting error, add enough components to the master mix for two additional reactions.
5. In the order listed in Table 14.2, add the final volume of each component to a sterile microcentrifuge tube. Once completed, mix the components gently and place on ice.
6. Add 22.50 μl of the PCR master mix to each tube (from step 4 above) and place on ice.
7. For amplification, add the appropriate volume (use 5 ng) of template DNA (extracted DNA from the buccal swabs and the human cell lines) to each reaction tube.

Note: If the DNA is stored in Tris-EDTA (TE) buffer, the volume of the DNA sample should not exceed 20% of the final volume since components of the buffer compromise PCR amplification efficiency and quality. This rule does not apply to DNA stored in sterile water.

8. For the positive control, pipet 2.5 μl (5 ng of human cell line DNA) into a 0.5 ml microcentrifuge tube containing 22.5 μl of the PCR master mix.
9. For the negative control, pipet 2.5 μl of sterile water (not template DNA) into a 0.5 ml microcentrifuge tube containing 22.5 μl of the PCR master mix.
10. Optional: Add one drop of mineral oil to each microcentrifuge tube to prevent evaporation. Close the tubes and centrifuge briefly (5 s). Depending on the thermal cycler model used, this step may be omitted.

TABLE 14.2
Multiplex Reactions Containing Three Loci

PCR Master Mix Component	Volume per Sample (μl)	Number of Reactions	Final Volume (μl)
Sterile water	17.35		
STR 10× buffer	2.50		
Multiplex 10× primer pair mix	2.50		
Taq DNA polymerase (at 5 u/μl)[a]	0.15 (0.75 u)		
Total volume	22.50		

[a] The volumes or values given for *Taq* DNA polymerase assume a concentration of 5 u/μl. If the final volume is less than 0.5 μl, the enzyme can be diluted with STR 1× buffer and then a larger volume added. Since the enzyme cannot be stored diluted, prepare only the amount that you will need. The amount of sterile water can be adjusted accordingly so that the final volume of each reaction is 25 μl.

11. Place the reaction tubes into a thermal cycler programmed to run at least 35 cycles with the following parameters:

Step 1:	2.0 min	94°C	Denaturation
Step 2:	1.0 min	94°C	Denaturation
Step 3:	1.0 min	64°C	Annealing
Step 4	1.5 min	70°C	Extension
Step 5:	Repeat steps 2–4 for a total of 35 times		
Step 6:	Soak	4°C	

Note: The parameters outlined above may vary according to the thermal cycler used for the PCR amplification.

12. Start or run the PCR incubation reaction. When the thermal cycler's program is completed (approximately 3 h), the tubes containing the PCR products will be removed by your instructor and stored at 0°C.

Visualization of the STR PCR Products

1. Because small PCR products (or DNA fragments) in the range of 150 to 400 base pairs are expected, a high concentration of agarose is required for adequate separation during gel electrophoresis. Prepare the agarose gel as described in the "Procedure" section of Exercise 9, step 1.
2. Remove 2.5 μl of the PCR product(s) and place into separate microcentrifuge tubes. Add 2.5 μl of STR 2× loading solution (supplied with kit) to each tube.
3. Add 2.5 μl (50 ng) of pGEM DNA markers (supplied with kit) to 2.5 μl of STR 2× loading solution.

Note: The pGEM DNA markers are visual standards used to confirm allelic size ranges for each locus. The markers consists of 15 DNA fragments with weights (bp) of 2,645, 1,605, 1,198, 676, 517, 460, 396, 350, 222, 179, 126, 75, 65, 51, and 36.

4. Add 2.5 μl of the STR allelic ladder (supplied in kit) to 2.5 μl of STR 2× loading solution for each ladder lane (at least two per gel).
5. Using a different pipette tip for each sample, load the DNA samples, mixed with loading solution (5 μl total volume), into the wells with the gel submerged. Be careful not to push the pipet tip through the bottom of the well in the gel.
6. Include on your gel the positive and negative controls. Load the gel as follows:

Lane 1:	pGEM markers	5 μl
Lane 2:	STR allelic ladder*	5 μl
Lane 3:	Human cell line DNA (positive control)	16 μl
Lane 4:	Negative control (no DNA)	16 μl
Lanes 5–11:	DNA from buccal swabs	16 μl
Lane 12:	STR allelic ladder	5 μl
Lanes 13–19:	DNA from buccal swabs	16 μl
Lane 20:	STR allelic ladder	5 μl

* For ease of interpretation, the allelic ladders can be run in lanes adjacent to each sample.

7. Set the voltage (100 V) and run the samples until the bromophenol blue tracking dye has moved 1–2 cm from the origin (i.e., well) or until the dye front is approximately 2 cm from the bottom of the gel. This should take less than 1 h.
8. Photograph the gel and determine the allele response and size at each locus. Direct comparison between the allelic ladders and amplified samples of the same locus should allow for the numerical assignment of each allele (see Figure 14.1 and Table 14.3).

Artifacts or unidentifiable DNA bands may be detected with this system. For questions related to these STR by-products refer to the troubleshooting guide in the technical manual for the GenePrint System.

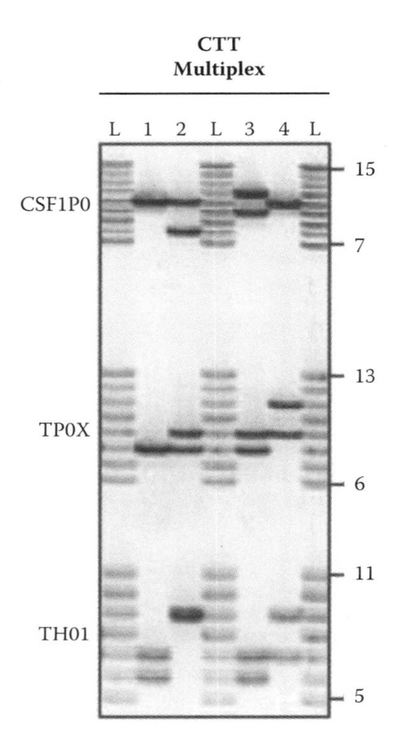

Figure 14.1
Profiles from the GenePrint STR Systems (CSF1PO, TPOX, TH01). Genomic DNA (lanes 1–4) was amplified using the CTT Multiplex STR System, separated in a 4% polyacrylamide denaturing gel, and detected using silver stain. The lanes labeled L contain the allelic ladders for each locus (i.e., CSF1PO, TPOX, and TH01). The numbers to the right of the image indicate the smallest to the largest number of repeats or numerical designation for each allele. (Courtesy of the Promega Corporation.)

TABLE 14.3
Locus-Specific Information for the GenePrint STR Systems (CSF1PO, TPOX, TH01)

Component Loci	Allelic Ladder Size Range (bases)	STR Ladder Alleles (number of repeats)	Other Known Alleles	K562 DNA Allele Sizes
CSF1PO	295–327	7, 8, 9, 10, 11, 12, 13, 14, 15	6	9, 10
TPOX	224–252	6, 7, 8, 9, 10, 11, 12, 13	None	8, 9
THO1	179–203	5, 6, 7, 8, 9, 10, 11	9.3	9.3, 9.3

<table>
<tr><th colspan="12">Sample Setup for Thermal Cycler</th></tr>
<tr><td colspan="4">Analyst:</td><td colspan="4">Date:</td><td colspan="4">Gel no.:</td></tr>
<tr><td colspan="6">Thermal cycler:</td><td colspan="6">Start time:</td></tr>
<tr><td colspan="6">Date of last calibration:</td><td colspan="6">System:</td></tr>
<tr><td>A1</td><td>A2</td><td>A3</td><td>A4</td><td>A5</td><td>A6</td><td>A7</td><td>A8</td><td>A9</td><td>A10</td><td>A11</td><td>A12</td></tr>
<tr><td>B1</td><td>B2</td><td>B3</td><td>B4</td><td>B5</td><td>B6</td><td>B7</td><td>B8</td><td>B9</td><td>B10</td><td>B11</td><td>B12</td></tr>
<tr><td>C1</td><td>C2</td><td>C3</td><td>C4</td><td>C5</td><td>C6</td><td>C7</td><td>C8</td><td>C9</td><td>C10</td><td>C11</td><td>C12</td></tr>
<tr><td>D1</td><td>D2</td><td>D3</td><td>D4</td><td>D5</td><td>D6</td><td>D7</td><td>D8</td><td>D9</td><td>D10</td><td>D11</td><td>D12</td></tr>
<tr><td>E1</td><td>E2</td><td>E3</td><td>E4</td><td>E5</td><td>E6</td><td>E7</td><td>E8</td><td>E9</td><td>E10</td><td>E11</td><td>E12</td></tr>
<tr><td>F1</td><td>F2</td><td>F3</td><td>F4</td><td>F5</td><td>F6</td><td>F7</td><td>F8</td><td>F9</td><td>F10</td><td>F11</td><td>F12</td></tr>
<tr><td>G1</td><td>G2</td><td>G3</td><td>G4</td><td>G5</td><td>G6</td><td>G7</td><td>G8</td><td>G9</td><td>G10</td><td>G11</td><td>G12</td></tr>
<tr><td>H1</td><td>H2</td><td>H3</td><td>H4</td><td>H5</td><td>H6</td><td>H7</td><td>H8</td><td>H9</td><td>H10</td><td>H11</td><td>H12</td></tr>
</table>

Results

STR Test Gel		
Analyst:	Lab number:	
Date:		
Well No.	Sample	DNA (ng)
1	pGEM marker	
2	STR allelic ladder	
3	Human cell line (positive control)	
4	No DNA (negative control)	
5		
6		
7		
8		
9		
10		
11		
12	STR allelic ladder	
13		
14		
15		
16		
17		
18		
19		
20	STR allelic ladder	

Reagents	Lot No.	Source
Agarose		
1× TBE (gel buffer)		
1× TBE (tank buffer)		
Loading buffer		
Ethidium bromide		
Visual marker		

Gel Electrophoresis		
Time on:	Voltage:	mAMPs
Time off:	Voltage:	mAMPs
Gel prepared by:	Date:	

Reporting Form

Tape your
Gel photo
Here

STR Test Gel		
Analyst:	Lab number:	
Date:		
Well No.	Sample	DNA (ng)
1		
2		
3		
4		
5		
6		
7		
8		
9		
10		
11		
12		
13		
14		
15		
16		
17		
18		
19		
20		

Interpreting Test Results

Your group will work together to interpret the photograph of your gel with the test results. Attach the gel photograph to the reporting form (see above). Determine the allele response and size at each locus. Direct comparison between the allelic ladders, designated as L, and amplified samples of the same locus should allow for the numerical assignment of each allele (see Figure 14.1 and Table 14.3).

For each lane, state the approximate size of each allele and if the person appears to be heterozygous or homozygous at the different STR loci. Also, examine the other groups' test results and attempt to determine the number of responses or the different alleles present in the class. Are any of the profiles similar? Are any two genotypes the same?

STR Typing Results			
	Locus		
Sample	CSF1PO	TPOX	THO1
1			
2			
3			
4			
5			
6			
7			
8			
9			
10			
11			
12			
13			
14			
15			
16			
17			
18			
19			
20			

In the forensic laboratory, the resulting STR profiles from the evidentiary samples are interpreted by direct comparison to DNA standards, allelic ladders, and the known reference standards (known DNA profiles from the victim and suspect). The analysis generally leads to three possible outcomes:

Match: DNA bands or peaks between the compared STR profiles have the same genotypes (or allelic profile), and no unexplainable differences exist between the samples. Statistical evaluation of the significance of the match is usually stated in the final report generated by the DNA analyst.

Exclusion: The genotype comparison shows differences in the allelic profiles that can only be explained by the two samples originating from different sources.

Inconclusive: The data do not support a conclusion as to whether the profiles match. This finding might be reported if two analysts remain in disagreement after review and discussion of the data and it is felt that insufficient information exists to support any conclusion.

Once a match is declared between the evidence sample and the reference sample, the question then becomes what is the strength of the inference or the weight of the evidence that the two profiles match. If many individuals in a given population share this profile, then the strength of the inference is minimal because there is a chance that someone else from that population may have contributed the same DNA profile. So the question becomes what is the probability of randomly selecting an individual from a given population that has the same allelic profile as observed in the evidence sample(s)? This question is answered by determining how many individuals in a population might possess the same allelic profile as observed in the evidentiary sample. Thus, the first step is to determine the allelic frequency at each locus tested by utilizing existing population databases (e.g., African American, Caucasian, and Southeast and Southwest Hispanic). The allelic frequencies are used to determine the frequency of a genotype at one locus with the calculations of homozygotes being derived differently than for heterozygote profiles. The resulting allelic frequencies from the different loci are multiplied together to obtain the frequency of the observed DNA profile in a given population. This value is usually expressed as the probability of randomly selecting an unrelated individual with a DNA profile matching the DNA developed from the evidence sample is 1 in greater than XXX (e.g., 1 in 1 million or 1 in 1 billion), where XXX represents the number of people that would need to be sampled or tested that would yield the same genotype or profile at least once. It should be noted that the final frequency or probability calculation is an estimate of the frequency of the evidence profile and not that of the suspect, unless, of course, the suspect's profile was used to determine the overall probability. Finally, the power of discrimination or the weight of the evidence increases with the number of loci tested. For example, in the above STR exercise the power of discrimination for the three loci tested is approximately 1 in 410; whereas if 16 loci were tested (i.e., Applied Biosystems PowerPlex® 16 System), the potential discriminatory power is 1 in 6.5 billion, which is approximately the world population.

Questions

1. What are the advantages of using STR systems versus some of the earlier DNA typing techniques?

2. STR loci chosen for use in the forensic community have many characteristics. Describe three favorable characteristics of STR loci.

3. What are some of the challenges that a forensic DNA analyst confronts with STR typing?

4. Why are STRs preferred genetic markers?

Chapter 15

Exercise 11

Using Short Tandem Repeat (STR) Analysis to Determine Paternity (A Case Study)

Introduction

DNA typing is the most accurate form of paternity testing possible. DNA typing can indicate with 100% certainty if the tested male is excluded as the biological father or will demonstrate with a high degree of scientific certainty (i.e., greater than 99.9% probability) if the tested male is the biological father. DNA paternity tests can be used to answer questions or issues related to

- Paternity/maternity identification and verification
- Child support and custody disputes
- Suspected incest cases
- Inconclusive paternity results from other methods
- Single-parent cases where paternity or maternity is in question
- Newborn testing
- Prenatal paternity cases
- Identification of father in surrogate mother cases
- Estate/trust disputes

Parentage testing is performed by collecting biological samples (e.g., blood or buccal swabs) from the mother, the child, and the alleged biological father. For newborns, testing can be performed using umbilical blood from the umbilical cord. In unusual circumstances, DNA can be collected from other sources, as previously described (see "Types of Biological Specimens, Chapter 2"). DNA testing is based on genetic information that is passed on from the parents to their children (see Table 15.1). In cases where the alleged father is unavailable for testing, partial pedigree analysis can be conducted using DNA samples from the parents of the alleged father. If necessary, siblings of the alleged father can also be used.

Objective

In this exercise, DNA has been extracted from samples (i.e., buccal swabs) collected from the mother, the child, and the alleged father. The isolated DNA was amplified at multiple STR loci using polymerase chain reaction (PCR) and the GenePrint STR Systems (Promega Corporation). Following amplification, the amplified STR products were separated and identified using polyacrylamide gel electrophoresis (PAGE). In

TABLE 15.1
PowerPlex 16 BIO Typing Results

Locus	15-Year-Old Mother	Child	18-Year-Old Boyfriend	Alleged Father
FGA	21, 23	22, 23	21, 22	20, 22
TPOX	8, 10	8, 11	11, 11	11, 11
D8S1179	12, 13	12, 13	13, 13	13, 13
vWA	14, 18	14, 15	17, 18	14, 15
Penta E	10, 11	11, 16	10, 11	14, 16
D18S51	13, 14	14, 14	14, 17	14, 17
D21S11	31.2, 32.2	30, 32.2	30, 30	29, 30
TH01	6, 9	6, 9.3	5, 6	6, 9.3
D3S1358	16, 17	16, 17	15, 15	15, 17
Penta D	11,14	11, 11	11, 14	11, 12
CSF1PO	8, 12	8, 12	8, 11	12, 12
D16S539	11, 13	9, 11	9, 13	9, 10
D7S820	9, 10	8, 9	8, 8	8, 9
D13S317	11, 12	10, 11	7, 12	10, 13
D5S818	7, 8	7, 11	11, 12	11, 12
Amelogenin	XX	XX	XY	XY

this exercise, you will be analyzing data, specific to a case study, generated using PAGE. As the amplified STR products are separated, the fluorescently labeled DNA molecules are excited by the laser light source, the emission captured by a detection system, and recorded as a chemilumigraph (e.g., bar code format) by the computer.

Equipment and Material

1. 0.5, 1.0, and 1.5 ml Eppendorf/microcentrifuge tubes
2. Agarose (DNA typing grade)
3. Mineral oil (optional)
4. 15 ml polypropylene test tube
5. Double-distilled water
6. Ice in buckets
7. Adjustable-volume digital micropipets (2–200 μl range)
8. Aerosol-resistant pipet tips
9. GenePrint STR Systems (PowerPlex 16 BIO System) (Promega Corporation, Madison, WI)
10. *Taq* DNA polymerase (not supplied in kit)
11. Disposable gloves
12. Genomic DNA (10 ng/μl) from mother, child, and alleged father
13. Molecular weight markers (526 to 22,621 bp)
14. Ethidium bromide/Coomassie blue
15. Bromophenol blue tracking dye
16. 125 ml Erlenmeyer flask
17. 1× Tri-Borate-EDTA (TBE) buffer (loading buffer)

18. 30 or 50 ml conical tubes
19. Incubator/water bath at 56°C
20. Electrophoresis systems (gel tray/combs)
21. Power pack/supply
22. DNA thermal cycler
23. Microcentrifuge
24. Microwave or hot plate

Procedure

To prevent cross-contamination, the use of disposable gloves and aerosol-resistant pipet tips is highly recommended.

1. Refer to Exercises 6 and 7 for the methods or steps used in the collection and concentration of cells and for methods outlining cell lysis and the collection of DNA.
2. Refer to Exercise 10 for the methods or steps used in setting up the PCR amplification and for the visualization of STR PCR products.

A Case Study

In 2013, a pregnant 15-year-old female claimed that her father had sexually assaulted her, which resulted in her pregnancy. The father, in his early fifties, was arrested and charged with incestrial pedophilia and sexual abuse. The father denied all claims and stated that his daughter was sexually active with her 18-year-old boyfriend. To establish a genetic profile of the fetus and to determine the biological father, a chorionic villus sample (CVS) from the fetus was collected. Samples (i.e., buccal swabs) were also collected from the mother (i.e., the 15-year-old daughter), the 18-year-old boyfriend, and the alleged father. All samples were sent to the state laboratory for DNA analysis. DNA was extracted from all samples, purified, and subjected to STR analysis using the PowerPlex 16 BIO Typing System (Promega Corporation). The fluorescently labeled STR products were then separated by gel electrophoresis and the DNA molecules captured using a fluorescent detection system. The STR typing results of the mother, the child, the 18-year-old boyfriend, and the alleged father are shown in Table 15.1.

Results

In Table 15.1, DNA was extracted from samples collected from the mother, the child, and the alleged father and amplified at the STR loci using the PowerPlex 16 BIO System.

Interpreting Test Results

Your group will work together to interpret the data generated from the STR analysis for the above-referenced case study. To determine if the alleged father or the 18-year-old boyfriend is the biological father, you should first determine which alleles (or STR fragments) of the child were donated by the mother. This analysis should be performed for each locus. Second, determine if the remaining allele at each locus of the child could have been contributed by the alleged father or the boyfriend. If one allele at any locus does not match, the alleged father or the boyfriend is excluded. If all of the remaining alleles at each locus match, then based on the alleged father's or the boyfriend's STR profile, he is presumed to be the biological father. On the basis of these results obtained from all genetic systems tested, the alleged father or the boyfriend cannot be excluded as the biological father of the child. The paternity probability, determined to be 99.99%, would also support these findings that the alleged father or the boyfriend is, in fact, the biological father.

Questions

1. Based on your analysis of the STR typing results, was the alleged father excluded as the biological father, or were his genetic markers consistent with those observed in the child and thus he could not be ruled out as the biological father?

2. Based on your analysis of the STR typing results, was the boyfriend excluded as the biological father, or were his genetic markers consistent with those observed in the child and thus he could not be ruled out as the biological father?

3. Using the same STR typing results would you be able to determine maternity in this instance? Why or why not?

4. Assuming one of the biological parents' profiles was unavailable, would you be able to determine the other parent's genetic contribution to the child? How?

5. Assuming that the alleged father (or the boyfriend) refused to provide a sample for DNA testing but you had legal access to his home, what samples would you collect for analysis? The idea is to collect enough material to generate a DNA profile.

6. If the child displayed homozygosity at one locus (e.g., for FGA: 22, 22), would you be able to assign each allele to either parent? Why or why not?

7. How accurate is DNA paternity testing? Are the results conclusive?

8. Will the DNA paternity test results stand up in court? Explain your answer.

9. The collection of cheek cells (e.g., a buccal swab) is often performed instead of collecting blood as a source of biological material for DNA paternity typing. Will the resulting DNA profile from the buccal swab be as accurate as one degenerated from blood cells? Explain your answer.

Chapter 16

Exercise 12

Polymerase Chain Reaction (PCR)-Based Tests: Y Chromosome Short Tandem Repeat (Y-STR) Analysis (A Case Study)

Introduction

The ability to designate whether a sample originated from a male or female contributor is extremely valuable in sexual assault cases as well as in other capital cases. The most popular method for sex typing is the amelogenin typing system since the DNA encoding gender can be amplified in conjunction with STR analysis. However, in some instances, STR and amelogenin analysis is not adequate when multiple males contribute to an evidentiary sample (e.g., a blood sample containing DNA from more than one male). Recently, Y-STR analysis has become available to the forensic community and has provided identification where STR analysis was not definitive.

Several genetic markers have been identified on the Y chromosome that are distinct from markers on the autosomes and are useful for human (male) identification (Table 16.1). The Y-STR markers are found on the noncoding region located on both arms (i.e., p and q) of the Y chromosome. The Y-STR markers produce a haplotype profile when amplified from male DNA. Such a profile simplifies the interpretation of a mixture containing a male and female contributor by eliminating the female contribution from the amplification profile. This also eliminates the need to separate semen and vaginal epithelial cells prior to analysis. The Y-STR markers are extremely valuable in sexual assault cases where samples contain multiple male contributors.

Y-STR markers are also useful in the analysis of lineage and the reconstruction of family relationships. In essence, a sample from a male may be compared with another male or his brother, father, paternal grandfather, or paternal uncles for identification purposes and familial relationships. Since these markers are only paternally inherited, they are useful in paternity-related matters. In addition, Y-STR markers' use and effectiveness in lineage studies can extend to answer questions of common ancestral geographical origin. Y-STR markers, together with mitochondrial DNA (mtDNA) markers (see Exercise 13), will complement each other in these ancestral analyses.

With Y-STR analysis, the mode of inheritance has statistical implications. Unlike STR markers that are on separate autosomes (chromosomes other than the sex chromosomes), are not linked, and therefore undergo independent assortment during gametogenesis, Y-STR markers are linked on the Y chromosome. Statistically, this means that the product rule can be applied to STR markers found on the autosomes, but not for Y-STR markers located on the Y chromosome. Instead, the frequency of the Y-STR profile is determined by the counting method, and the statistical power relies on the size of the database. For example, a Y-STR profile might be observed once in a database of 4,500 profiles and reported as 1 out of 4,500 profiles. Or, a profile may not be observed at all in the database and is reported as 0 out of 4,500 profiles. It should

TABLE 16.1
Y-STR Loci of the PowerPlex Y System

Y-STR Locus	Sequence
DYS391	TCTA
DYS389I	[TCTG][TCTA]
DYS439	GATA
DYS389II	[TCTG][TCTA]
DYS393	AGAT
DYS390	[TCTG][TCTA]
DYS385a/b	GAAA
DYS438	TTTTC
DYS437	[TCTA][TCTG]
DYS19	TAGA
DYS392	TAT

be noted that since all paternity-linked males have the same Y-STR profile, brothers would have to be differentiated using STR analysis.

Objective

In this exercise, DNA has been extracted from male buccal swabs or from a human cell line of male origin, incubated with appropriate PCR reagents, and multiple Y-STR loci amplified using the PowerPlex Y System (Promega Corporation). Following PCR, the amplified Y-STR products can be separated and analyzed using polyacrylamide gel electrophoresis or capillary electrophoresis (Figure 16.1). In this exercise, you will be analyzing data, specific to a case study, generated using capillary electrophoresis (CE). As the amplified Y-STRs products are separated, the fluorescently labeled DNA molecules are excited by the laser light source, the emission captured by a detection system, and recorded as an electropherogram by the computer.

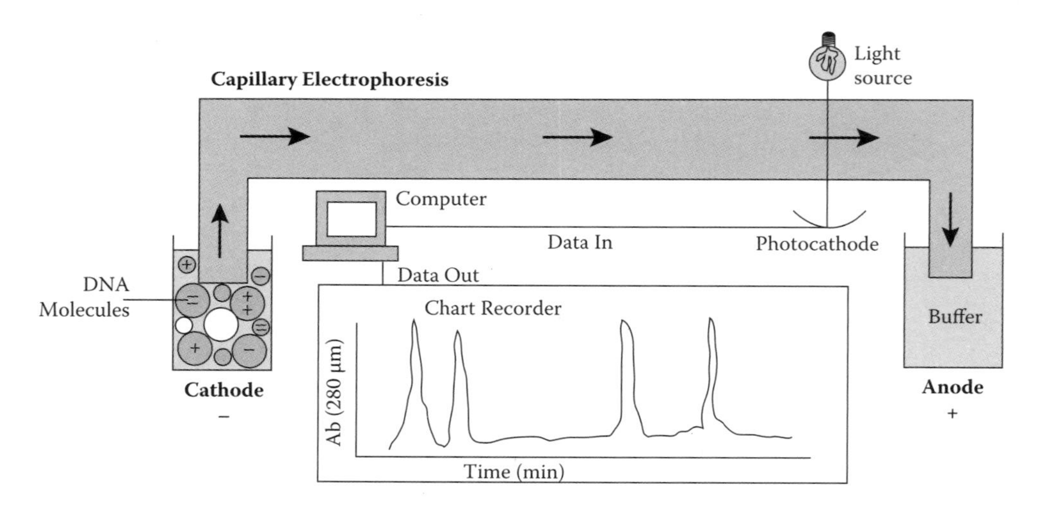

Figure 16.1
Schematic of the capillary electrophoresis system. Samples are injected into the tube on the left (cathode) and travel to the right or to the anode. The fluorescently labeled DNA molecules (or amplified products) are excited by the light source, captured by a detection system, and recorded as an electropherogram by the computer. (Courtesy of www.CEandCEC.com.)

Equipment and Material

1. 0.5, 1.0, and 1.5 ml Eppendorf/microcentrifuge tubes
2. 30 or 50 ml conical tubes
3. Mineral oil (optional)
4. 15 ml polypropylene test tube
5. Double-distilled water
6. Ice in buckets
7. Adjustable-volume digital micropipets (2–200 µl range)
8. Aerosol-resistant pipet tips
9. PowerPlex Y Systems (Promega Corporation, Madison, Wisconsin)
10. *Taq* DNA polymerase (not supplied in kit)
11. Disposable gloves
12. Agarose (DNA typing grade)
13. Precast polyacrylamide gels
14. Genomic DNA from human cell lines
    a. HEP G2—hepatocellular carcinoma (liver), male (10 ng/µl)
15. Molecular weight markers (526 to 22,621 bp)
16. Ethidium bromide/Coomassie blue
17. Bromophenol blue tracking dye
18. 125 ml Erlenmeyer flask
19. 1× Tris-Borate-EDTA (TBE) buffer
20. Incubator/water bath at 56°C
21. Electrophoresis systems (gel tray/combs)
22. Power pack/supply
23. Microwave or hot plate
24. DNA thermal cycler
25. Microcentrifuge
26. Tabletop clinical centrifuge

Procedure

To prevent cross-contamination, the use of disposable gloves and aerosol-resistant pipet tips is highly recommended. A helpful organizational sheet is provided at the end of the exercise to record your Y-STR typing results.

1. Refer to Exercises 6 and 7 for the methods or steps used in the collection and concentration of cells and for methods outlining cell lysis and the collection of DNA.
2. Refer to Exercise 10 for setting up the PCR amplification reaction. Keep all samples and reagents on ice.
3. Determine the number of reactions to be set up. Positive and negative controls should also be included when determining the number of reactions.
4. For each reaction, label one sterile 0.5 ml microcentrifuge tube and place into a rack.

TABLE 16.2
Master Mix for the PowerPlexY System

PCR Master Mix Component	Volume per Sample (µl)	Number of Reactions	Final Volume (µl)
Sterile nuclease-free water			
Gold ST*R 10× buffer	2.50		
PowerPlex Y 10× primer pair mix	2.50		
AmpliTaq Gold DNA polymerase (at 5 u/µl)[a]	0.55 (2.75u)		
Total volume	25.00		

[a] Template DNA volume (0.25–1 ng)—up to 19.45 µl.

5. To determine the final volume of the master mix needed for all of the reactions calculate the required amount of each component of the PCR master mix (see Table 16.2). Multiply the volume (µl) per sample by the total number of reactions (from step 2) to obtain the final volume (µl).
6. The reaction tubes were placed in a thermal cycler programmed to run in two phases: 10 cycles at set parameters followed by 22 cycles:

   Step 1: 11.0 min 95°C
   Step 2: 1.0 min 96°C
   Step 3: 1.0 min 94°C
   Step 4: 1.0 min 60°C
   Step 5: 1.5 min 70°C

   For 10 cycles, then:

   Step 6: 1.0 min 90°C
   Step 7: 1.0 min 58°C
   Step 8: 1.5 min 70°C

   For 22 cycles, then:

   Step 9: 30 min 60°C
   Step 10: Soak 4°C

Note: The parameters outlined above may vary according to the thermal cycler used for the PCR amplification.

Detection of the Y-STR PCR Products

1. Following PCR amplification, the fluorescently labeled Y-STR alleles are separated and sized using polyacrylamide gel electrophoresis (for PAGE) or capillary electrophoresis (for CE).
2. Samples are denatured by heating at 95°C for 3 min (for PAGE) or by diluting with a denaturant solution (for CE), then immediately chilled on ice.
3. The samples are loaded onto a gel (for PAGE) analysis or introduced into the capillary (for CE) analysis by injection. For PAGE analysis, multiple samples are separated and analyzed in 2.5 to 3 h. For CE analysis, only one sample is injected into a capillary tube for separation and analysis; however, this process is completed in a matter of a few minutes.

4. Detection of the sample analyzed by PAGE is performed by scanning each lane and then imaged using a computer detection system. Detection of the sample is performed automatically by the CE instruments by measuring the time span from injection to sample detection with a laser near the end of the capillary. In both instances (i.e., PAGE and CE), the laser excites the fluorescently labeled DNA fragments, which causes a fluorescent light emission. This emission is captured by the detection system and plotted as a function of the relative fluorescence intensity observed from each fluorescent dye attached to the DNA molecule. These signals, recorded as bands on a gel or as an electropherogram for CE, can then be used to detect and quantify the Y-STR PCR products (Figure 16.2).

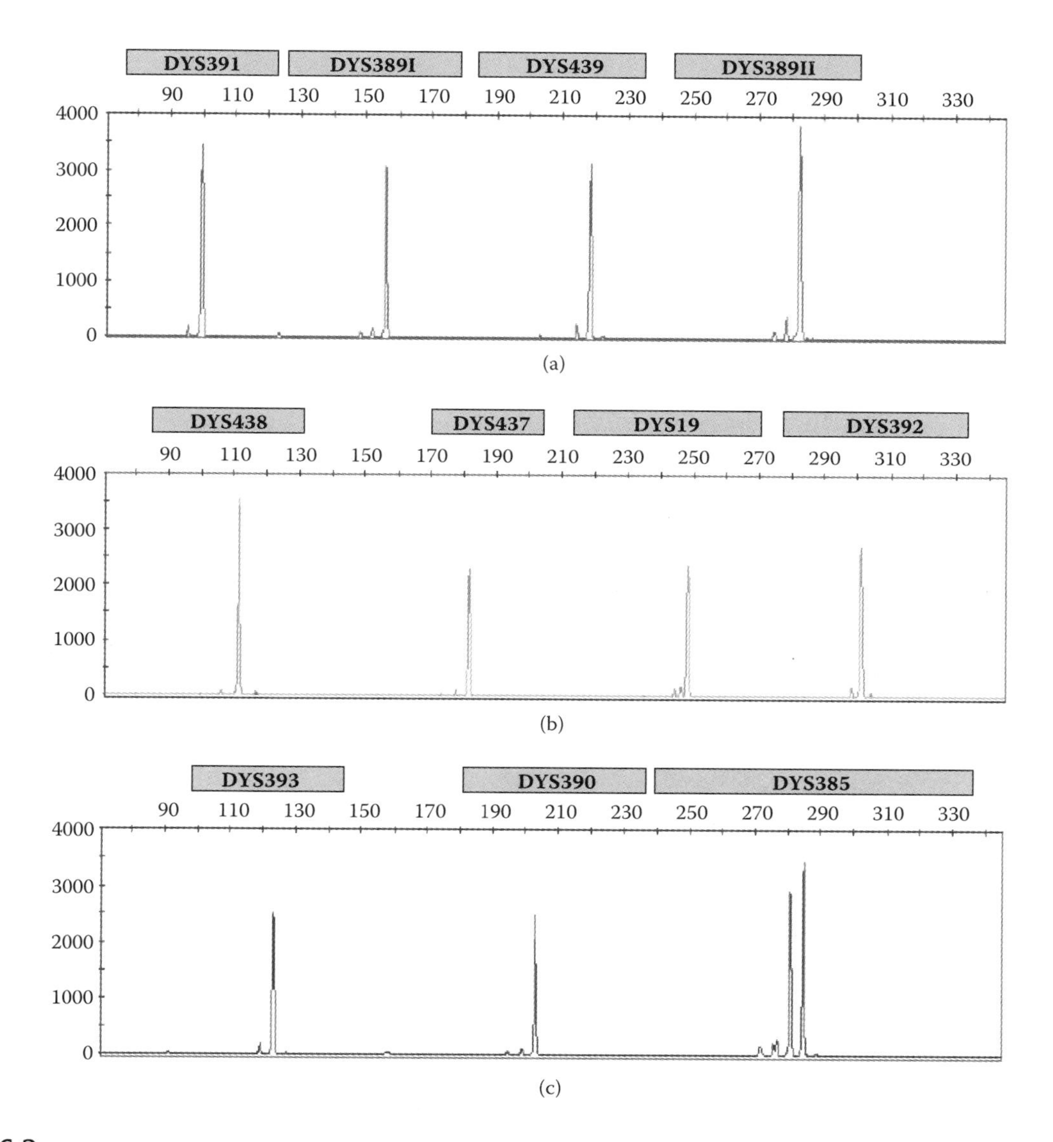

Figure 16.2
The PowerPlex Y System. A single-source sample from a male contributor was amplified using the PowerPlex Y System. The amplified products were captured using the Applied Biosystems 3130 Genetic Analyzer and analyzed using the GeneMapper ID software to generate the Y-STR profile. (a) An electropherogram of the DYS391, DYS389I, DYS439, and DYS389II loci. (b) An electropherogram of DYS438, DYS437, DYS19, and DYS392. (c) An electropherogram of DYS393, DYS390, and DYS385. (Courtesy of the Promega Corporation.)

A Case Study

A woman was walking to her car, which was parked in an underground parking garage. As she was unlocking her car door, a man approached her from behind, forced her into the back seat of her car, and raped her. After the attacker fled the scene, she immediately called the police from her cell phone. The police took her to the local hospital, where she was examined by a sexual assault nurse examiner. Vaginal swabs and a reference sample were collected and sent to the state's forensic laboratory for analysis. The woman had described her attacker as a tall and thin African American male with a tattoo on his right hand (the only part of him that was visible to her during the attack). From that description and a file of known sexual offenders, the police arrested a male suspect. A blood sample was collected from the suspect and sent to the forensic laboratory for analysis. Since the woman was not married, did not have a boyfriend, and had not had consensual sex in several weeks, samples from consensual partners were not needed for analysis.

At the laboratory, the samples were subject to DNA typing, specifically Y-STR analysis. The vaginal swabs taken from the victim were found to contain sperm and her own cells. The sperm cells were first separated from the victim's epithelial cells and the DNA isolated using differential extraction. The purified DNA was amplified by PCR and analyzed at 11 Y-STR loci (DYS391, DYS389I, DYS439, DYS389II, DYS438, DYS437, DYS19, DYS392, DYS393, DYS390, and DYS385). The suspect's DNA was also analyzed at the same Y-STR loci.

Data Analysis

1. Following amplification, the amplified Y-STR products from the suspect's known reference sample (see Table 16.3 for Y-STR typing results) and the sperm fraction from the vaginal swab were separated by CE (Figure 16.3).
2. The suspect's allelic response at each Y-STR locus can be identified by direct comparison between the Y-STR profile from the sperm fraction (Figure 16.3), the allelic ladders (Figure 16.4), and the electropherogram of the identified alleles (Figure 16.5). The allelic responses observed for each locus should allow for the numerical assignment of each allele. The electropherogram of the negative control should be devoid of any amplification products (results not shown). Conversely, the electropherogram of the positive control should consist of the male DNA standard with known allelic responses (data not shown).

TABLE 16.3
Y-STR Typing Results

Y-STR Loci	Suspect's Reference Sample
DYS391	10
DYS389I	14
DYS439	11
DYS389II	32
DYS438	10
DYS437	14
DYS19	15
DYS392	12
DYS393	14
DYS390	23
DYS385	15, 17

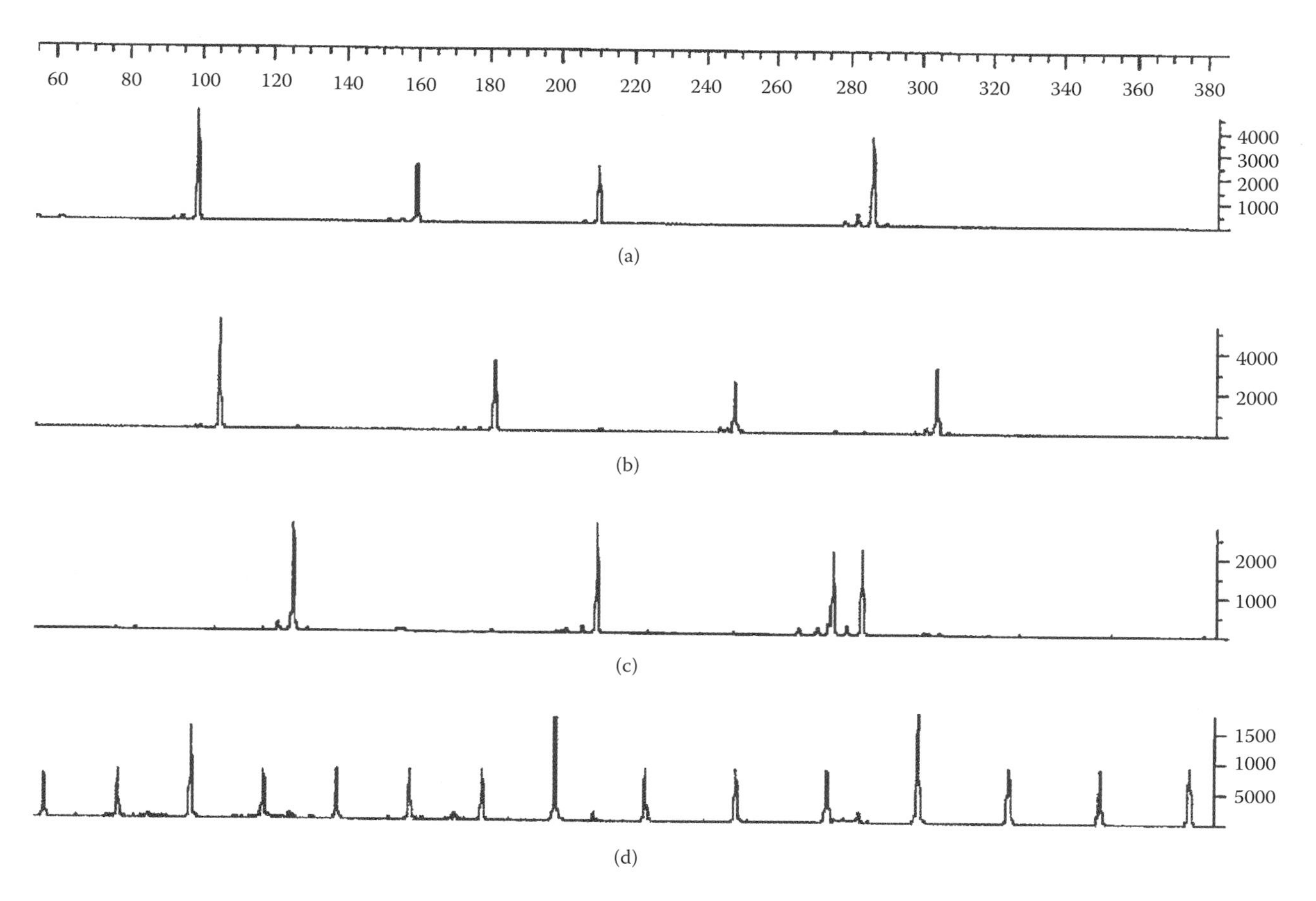

Figure 16.3
The PowerPlex Y-STR typing results. The Y-STR profile, shown in the electropherogram above, was obtained from the sperm fraction of the vaginal swab collected from the victim. Eleven Y-STR loci, specific to the male chromosome, were amplified and separated by capillary electrophoresis. (a) An electropherogram of the DYS391, DYS389I, DYS439, and DYS389II loci. (b) An electropherogram of DYS438, DYS437, DYS19, and DYS392. (c) An electropherogram of DYS393, DYS390, and DYS385. (d) An electropherogram showing the fragments of the internal lane standard.

Interpreting Test Results

Your group will work together to interpret the electropherograms. Using the Y-STR typing results from the suspect's reference sample (Table 16.3), the Y-STR profile from the sperm fraction of the vaginal swab from the victim (Figure 16.3) generated by CE, and the allelic ladder (Figure 16.4), determine the allelic numerical designation for each locus analyzed. In addition, determine the overall Y-STR profile from the sperm fraction and record your observation in the table below.

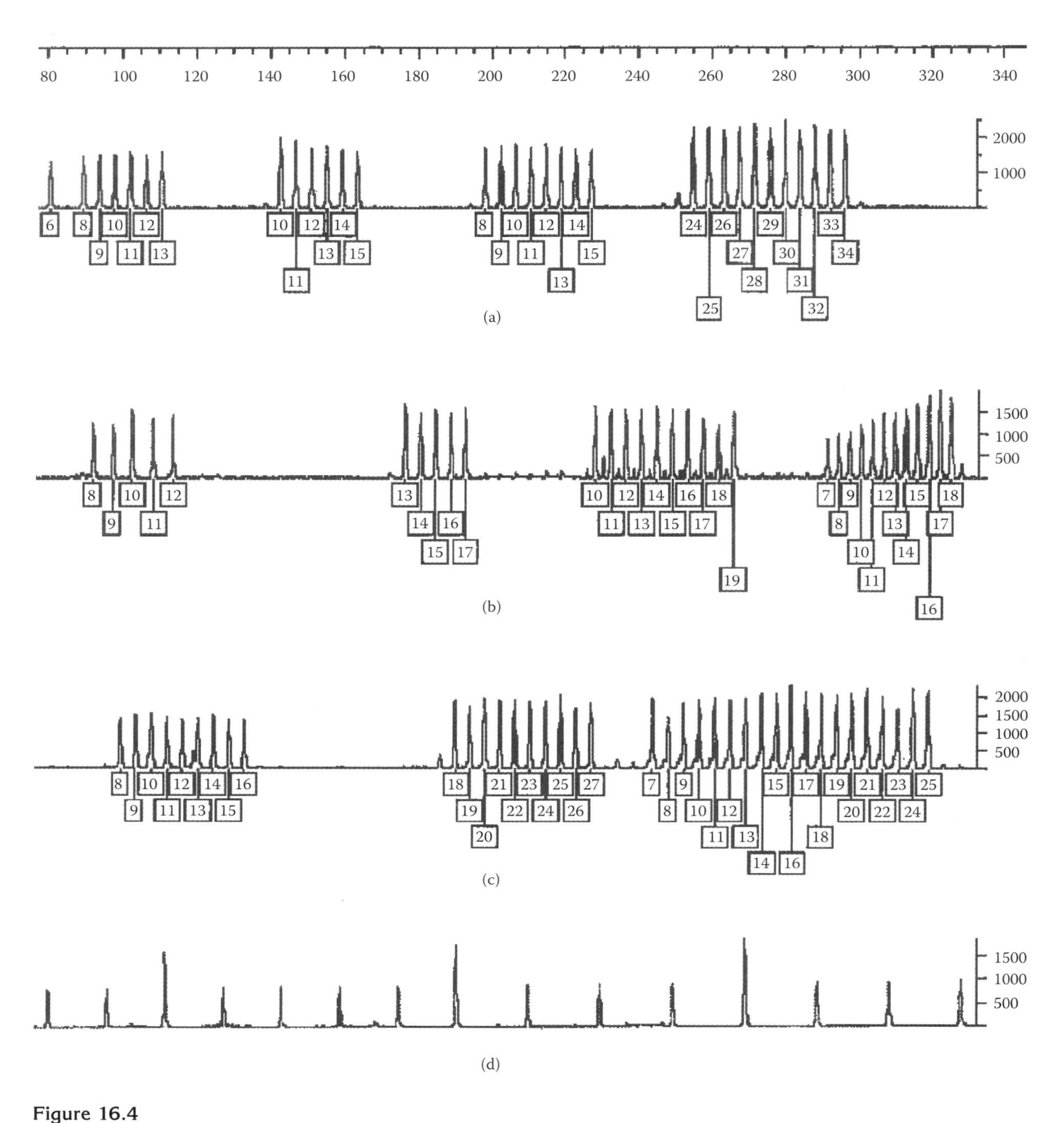

Figure 16.4
The PowerPlex Y allelic ladder mix. The allelic components and their allelic or numerical designations. (a) An electropherogram of the DYS391, DYS389I, DYS439, and DYS389II loci. (b) An electropherogram of DYS438, DYS437, DYS19, and DYS392. (c) An electropherogram of DYS393, DYS390, and DYS385. (d) An electropherogram showing the fragments of the internal lane standard.

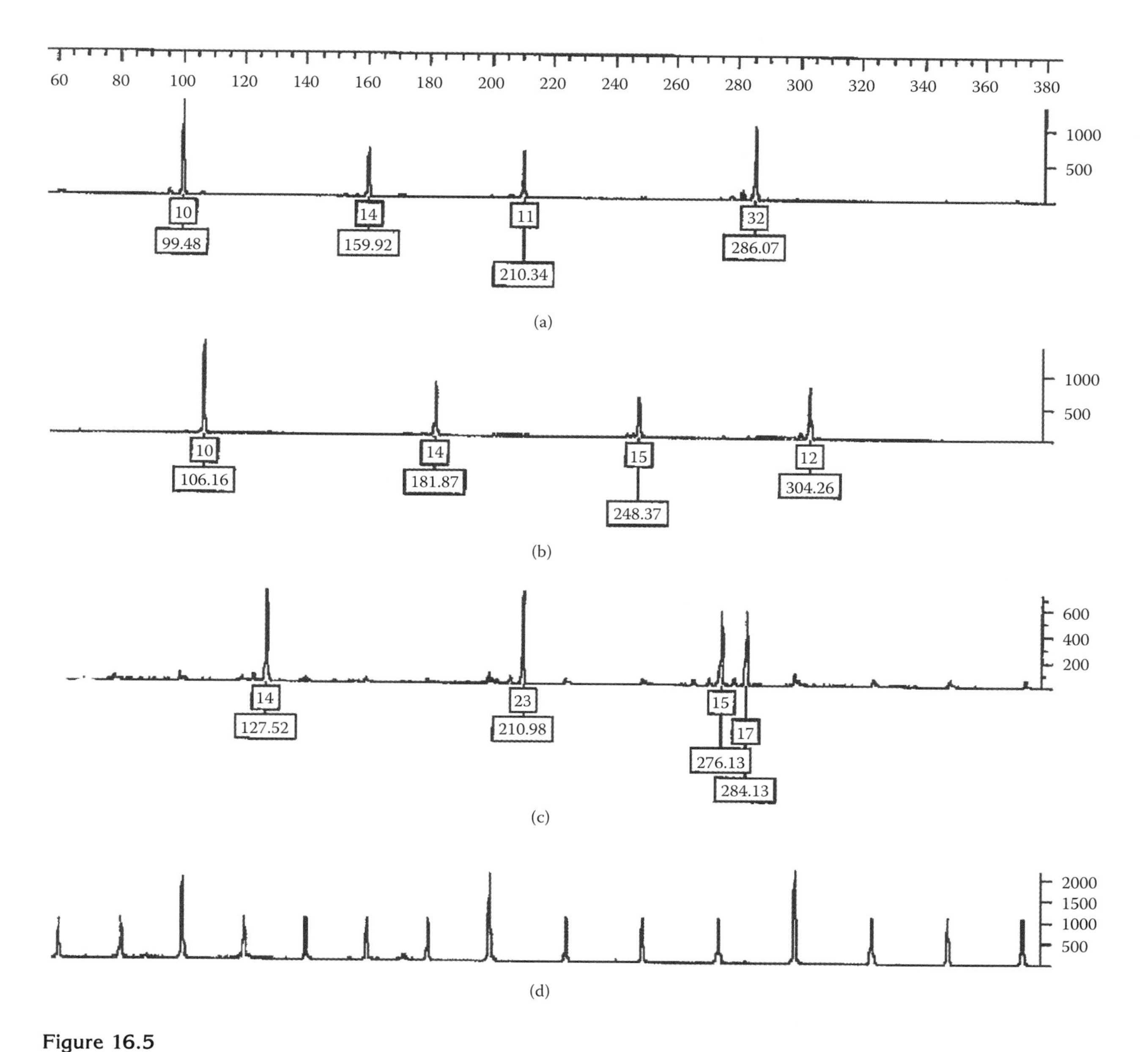

Figure 16.5
The PowerPlex Y-STR typing results: Alleles identified. The Y-STR profile, shown in the electropherogram, was obtained following amplification of the STR loci specific to the male chromosome. Direct comparison between the allelic ladder (Figure 16.4) and the amplified sperm fraction of the same locus (Figure 16.3) allowed for the numerical assignment of each allele. (a) An electropherogram of the DYS391, DYS389I, DYS439, and DYS389II loci. (b) An electropherogram of DYS438, DYS437, DYS19, and DYS392. (c) An electropherogram of DYS393, DYS390, and DYS385. (d) An electropherogram showing the fragments of the internal lane standard.

Y-STR Typing Results		
Y-STR Loci	Sperm Fraction from Vaginal Swab	Suspect's Reference Sample
DYS391		10
DYS389I		14
DYS439		11
DYS389II		32
DYS438		10
DYS437		14
DYS19		15
DYS392		12
DYS393		14
DYS390		23
DYS385		15, 17

Questions

1. Are the Y-STR profiles similar between the sperm fraction and the suspect's known reference sample? Explain your answer.

2. In this case study, the sperm cells were separated from the victim's epithelial cells at the start of the Y-STR analysis. Was this step necessary? Why or why not?

3. How does the Y-STR profile, generated by your group from Figure 16.3, compare to the Y-STR profile observed in Figure 16.5?

4. If multiple males were involved in the sexual assault, how would you differentiate between each contributor using Y-STR analysis? What type of Y-STR allelic response would you expect to see at each locus?

5. Would STR analysis complement the Y-STR results? Explain your answer.

Chapter 17

Exercise 13
Mitochondrial DNA (mtDNA) Analysis

Introduction

Mitochondrial DNA (mtDNA) typing is increasingly used in human identity testing when biological evidence may be degraded, when quantities of the samples in question are limited, or when nuclear DNA typing is not an option. Forensically relevant biological sources of mtDNA include, but are not limited to, hairs, bones, and teeth. In humans, mtDNA is inherited strictly from the mother. Consequently, mtDNA analysis cannot discriminate between maternally related individuals (e.g., mother and daughter, brother and sister). However, this unique characteristic of mtDNA is beneficial for missing person cases when mtDNA samples can be compared to samples provided by a maternal relative of the missing person.

In humans, the mtDNA genome is approximately 16,569 bases (A, T, G, and C) in length, containing a control region with two highly polymorphic regions (Figure 17.1). These two regions, termed hypervariable region 1 (HV1) and hypervariable region 2 (HV2), are 342 and 268 base pairs (bp) in length, respectively, and are highly variable within the human population. This sequence (the specific order of bases along a DNA strand) variability in either region provides an attractive target for forensic identification studies. Moreover, since human cells contain several hundred copies of mtDNA, substantially more template DNA is available for amplification using polymerase chain reaction (PCR) than nuclear DNA.

Mitochondrial DNA typing begins with the extraction of mtDNA from the mitochondria of human cells followed by PCR amplification of the hypervariable regions. The amplified mtDNA is purified, subjected to the dideoxy terminator method of sequencing (Sanger et al., 1977), with the final products containing a fluorescently labeled base at the end position. The products from the sequencing reaction are separated, based on their length, by gel electrophoresis. The resulting sequences or profiles are then compared to sequences of a known reference sample to determine differences and similarities between samples. Samples are not excluded as originating from the same source if each base (A, T, G, or C) at every position along the hypervariable regions is similar. This sequence, if determined to be similar between a known reference sample and an evidentiary sample, can be entered and searched in a database containing mtDNA sequences from four main racial groups (Caucasians, African Americans, Hispanics, and Asians). The search will generate a number that represents the number of observations of that sequence in each racial subgroup within the database. For example, a sequence might be seen three times in the database samples of Hispanic descent and not appear in the remaining database subgroups. Or, a sequence may not be observed at all in the database and is reported as not being observed. This number is usually reported as 1 out of 4,800 sequences or 0 out of 4,200 sequences. However, due to the size of the mtDNA database and to the unknown number of mtDNA sequences in the human population, a reliable frequency estimate is not provided. Consequently, mtDNA sequencing is becoming known as an exclusionary tool as well as a technique to complement other human identification techniques.

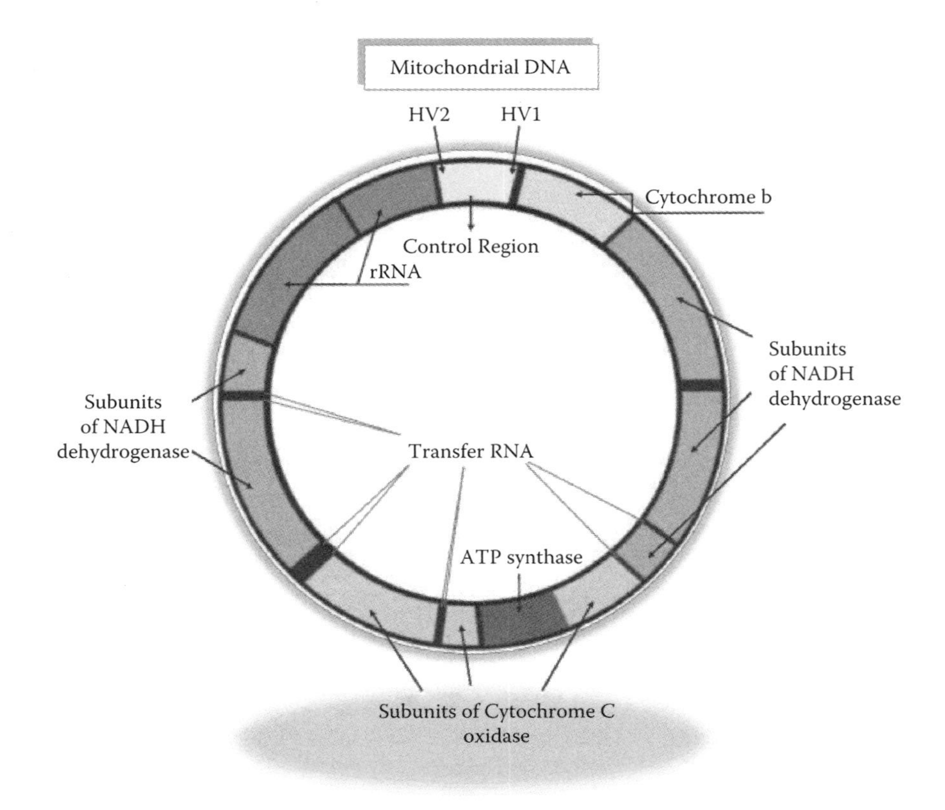

Figure 17.1
The mitochondrial genome in humans. The two noncoding hypervariable control regions (HV1 and HV2), located within the D-loop of the mtDNA genome, are positioned at the top of the figure. Each of the hypervariable regions is approximately 300 bp in length. The HV1 extends from nucleotide 16024 to 16383, and HV2 from nucleotide 57 to 372. (Courtesy of the University of Leicester, East Midlands Forensic Pathology Unit, Leicester, UK. http://www2.le.ac.uk/department/emfpu/genetics/explained/mitochondrial.)

Objective

In this exercise, you will isolate mitochondria from cells, extract mtDNA from the mitochondrial fraction, incubate the isolated mtDNA with appropriate PCR reagents, and amplify the hyperviable regions. The amplified mtDNA is then purified, subjected to sequencing, and separated by gel electrophoresis. You will compare the sequencing data from your known reference and unknown sample to a known standard.

Equipment and Material

1. 0.5, 1.0, and 1.5 ml Eppendorf/microcentrifuge tubes
2. 30 or 50 ml conical tube
3. Mineral oil (optional)
4. 15 ml polypropylene test tube
5. Deionized water
6. 10% ammonium persulfate solution
7. Adjustable-volume digital micropipets (2–200 μl range)

8. Aerosol-resistant pipet tips
9. *Taq* DNA polymerase
10. Disposable gloves
11. Ice in buckets
12. mtDNA from human cell lines
    a. HEP G2—an abbrviation for the cell line of a hepatocellular carcinoma (liver) cell line of male origin
13. Molecular weight markers (526 to 22,621 bp)
14. mtDNA primers
15. Ethidium bromide/Coomassie blue
16. Bromophenol blue tracking dye
17. TEMED (N, N, N′, N′-tetramethylethylenediamine)
18. Universal Serial Bus (USB) Thermo Sequenase Cycle Sequencing Kit (USB Corporation, Cleveland, Ohio)
19. 20 and 60 cc syringe/14-gauge needle
20. LI-COR infrared DNA analyzer (Model 4300) or equivalent
21. Microcentrifuge
22. DNA thermal cycler
23. Microwave or hot plate
24. Incubator/water bath at 56°C, 92°C

Procedure

To prevent cross-contamination, the use of disposable gloves and aerosol-resistant pipet tips is highly recommended. A helpful organizational sheet is provided at the end of the exercise to record your mtDNA typing data.

The protocol provided below is a basic guide to DNA sequencing using the LI-COR infrared DNA analyzer (Model 4300). Various aspects of sequencing are discussed (i.e., mtDNA analysis), including template preparation, primers used, reagents, labeled primer sequencing, gel preparation, and data analysis and interpretation.

The LI-COR system detects DNA using infrared (IR) fluorescence. In the dideoxy sequencing reaction, the DNA polymerase incorporates either a nucleotide or a primer labeled with an IRDye™ into a newly synthesized set of chain-terminated complementary strands. The IRDye-labeled fragments are separated by gel electrophoresis and are detected using a laser, which excites the dye on the DNA fragments. The emission or signal is a series of bands displayed on a computer in a "bar code" format similar to an autoradiograph. The bar code image is captured by the DNA sequencer and analyzed using specific software (e.g., e-Seq). The sequence data are determined for each lane, and the specific order of bases (A, T, G, and C) is determined. The sequence data are presented as a standard chromatogram or as ASCII text.

Template Preparation

The quality or purity of the DNA template will dictate the sequence data quality. Several DNA extractions and purification methods were described in previous exercises that will ensure and maximize the quality/purity of the DNA template (see Exercises 6–9). In addition to the quality of the DNA, it is important to determine the quantity of DNA in the known or evidentiary sample to be analyzed. Similar amounts of DNA template in each reaction will provide consistent data and similar band intensities. Template DNA concentration should

TABLE 17.1
Template Amount Used in Simultaneous Bidirectional Sequencing (SBS)

Size (bp)	Template (fmol)
300–600	50–100
600–1,200	125–225
1,300–1,800	250–300
>1,800	300–500

TABLE 17.2
Concentration of Template DNA Recommended for Labeled Primer Cycle Sequencing

Template	Amount (fmol)
Plasmid	200–500
PCR products	20–50
M13	100–200
Cosmids	50 (1.5 μg)

range between 0.5 and 1.0 μg/μl (see Tables 17.1 and 17.2). The amount of template DNA used in the reaction is based on the size of the DNA sequence between the two primers. If lower yields are obtained, concentrate the DNA by resuspending the final pellet in a smaller volume of buffer (see Exercise 7).

Primers

Successful sequencing reactions depend on many factors; however, primer design (i.e., 40–50% GC content; contain a G or C at the 3′ end, and avoid base repeats greater than three bases), primer purity and concentration are the most critical factors. Many primers are commercially available from several distributors and are, in general, of a high quality. If not commercially available, it is essential that the main impurities (e.g., salts or organic groups) that can affect the sequencing reaction have been removed.

Labeled Primer Sequencing

Simultaneous bidirectional sequencing (SBS) uses two labeled primers (e.g., forward and reverse primer pairs labeled with different IR dyes) on a single DNA template in a single reaction. An SBS reaction uses equal amounts of labeled primers (IRDye 700 and IRDye 800) to obtain equal signal strength in both channels.

Note: The Model 4300 detection system uses two separate lasers and detection that maximizes sequence accuracy (see Figure 17.2).

Setting Up the PCR Amplification

1. Label four 0.5 ml microcentrifuge tubes as follows: A, T, G, and C.
2. Prepare the following template-primer master mix (see Table 17.3). Add the largest volume first, and then add solutions in descending order based on volume. Mix the components by pipetting.
3. Add 4 μl of the template-primer master mix to each tube labeled A, T, G, and C.

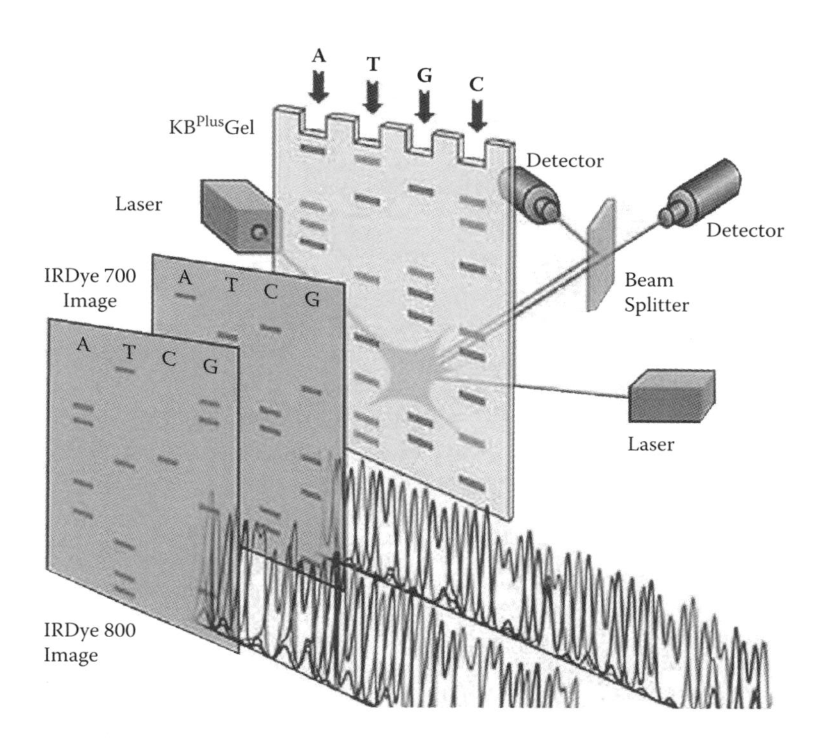

Figure 17.2
The LI-COR infrared DNA analyzer (Model 4300). Samples are loaded on the polyacrylamide gel and separated by electrophoresis. As the samples pass in front of the scanning laser/microscope, two photodiodes (e.g., the detectors) detect fluorescence. Each detector measures fluorescence from only one of the infrared dyes. A separate image (similar to an autoradiogram) for each IR dye is collected in real time and can be displayed in an Internet browser or LI-COR application software (i.e., e-Seq, Saga, etc.). (Used with permission from LI-COR Biosciences.)

TABLE 17.3
Template-Primer Master Mix

Template-Primer Master Mix Components	Volume
Template DNA	0.3 μl (300 ng)
IRDye 700 forward primer (1.0 pmol/μl)	1.5 μl
IRDye 800 reverse primer (1.0 pmol/μl)	1.5 μl
Thermo Sequenase reaction buffer	2.0 μl
2.5 mM dNTP nucleotide mix	1.0 μl
Thermo Sequenase DNA polymerase	2.0 μl
ddH_2O to bring final volume to 17.0 μl	__ μl
Total volume	17.0 μl

4. Add 4 μl of the A reagent to the tube labeled A, the T reagent to the tube labeled T, etc. (reagents supplied with Sequenase Cycle Sequencing Kit).
5. Add one drop of mineral oil to each microcentrifuge tube to prevent evaporation.

This step is required for thermal cyclers without heated lids. Close the tubes and centrifuge briefly (5 s).

6. Place the reaction tubes into a thermal cycler programmed to run at least 30 cycles with the following parameters:

   Step 1: 2.0 min 92°C Denaturation

   Step 2: 30 s 92°C Denaturation

   Step 3: 30 s 54°C Annealing

   Step 4: 1.0 min 70°C Extension

   Step 5: Repeat steps 2–4 for a total of 30 cycles

   Step 6: Soak or hold 4°C

Note: The parameters outlined above may vary according to the thermal cycler used for the PCR amplification.

7. Start or run the PCR incubation reaction.
8. At the completion of the cycling program add 4 µl of the IR^2 stop solution to each tube.
9. If mineral oil was used, remove the oil from each sample. Denature the samples at 92°C for 3 min. Then place the samples on ice.

Gel Electrophoresis

Assembling the Gel Apparatus

Follow the manufacturer's manual and protocol for specific instructions on assembling the electrophoresis apparatus, preparing the gel, pouring the gel, preparing pre-electrophoresis, starting the run, using the e-Seq software, and disassembling and cleaning up the gel apparatus. The protocol outlined below highlights the major steps in the mtDNA sequencing analysis.

1. Assemble the gel sandwich by laying the back plate (Figure 17.3, #6) down on the bench (gel side up) and placing two spacers (Figure 17.3, #5) along the edges of the long axis of the glass plate.
2. Place the front plate (Figure 17.3, #7; gel side down) on top of the bottom plate containing the spacers, making sure the plates are aligned at the bottom.
3. Place the left and right rail assemblies over the long axis of the plate edges (Figure 17.3, #8 and #9). Tighten the glass clamp knobs on each rail finger tight.

Gel Preparation

1. The gel and running buffer solutions are prepared from a 10× Tris-Borate-EDTA (TBE) buffer. Empty the contents of the KB^{Plus}10× TBE package (supplied with sequencing kit) in a 1 L beaker and add distilled water to bring the volume up to 800 ml. Stir the solution until all of the solids have gone into solution.
2. Bring the final volume to 1 L with distilled water. Store at room temperature.
3. Prepare the running buffer (0.8×) by adding 80 ml of the 10× TBE to 920 ml of distilled water and mix well.
4. For polymerization of the gel, an ammonium persulfate solution (APS) is prepared by adding 0.1 g of ammonium persulfate to 1.0 ml of deionized water. The APS should be prepared fresh.
5. Bring 40 ml of the KB^{Plus} Gel Matrix to room temperature.
6. Add 175 µl of the 10% APS and 17.5 µl of TEMED to the 40 ml KB^{Plus} Gel Matrix and mix thoroughly.
7. Using a 60 cc syringe with a 14-gauge needle, draw the gel solution (from step 6 above) into the syringe, and inject the solution into the gel cassette.

Note: The KB^{Plus} Gel Matrix is a ready-to-use solution containing polyacrylamide. The final gel concentration is 3.7%, that is, 66 cm long and 0.2 mm thick. There are other commercial acrylamides available that can be used in the LI-COR system.

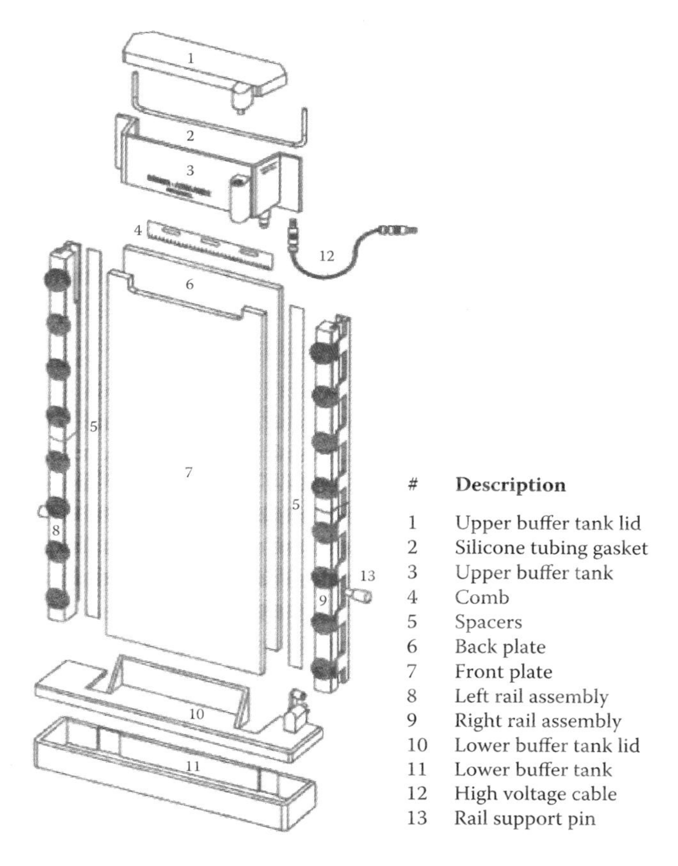

Figure 17.3
Expanded view and list of parts for the gel apparatus. (Used with permission from LI-COR Biosciences, Lincoln, Nebraska.)

8. After pouring the gel, invert the sharkstooth comb with the teeth down (Figure 17.3, #4) and insert upside down at the top of the gel cassette between the front and back plates (Figure 17.3, #6 and #7).
9. The sharkstooth comb (Figure 17.4) is inverted prior to polymerization of the gel to make a trough. After polymerization, the comb is removed, inverted teeth down, and inserted into the gel, forming the wells for the samples.
10. Place the casting plate (part not shown) at the top of the gel cassette and on the front plate. The casting plate will secure the comb until polymerization has occurred.
11. Allow at least 1.5 h for polymerization.

Electrophoresis Preparation

1. After polymerization, remove the casting plate and the sharkstooth comb.
2. Place the silicone tubing gasket (Figure 17.3, #2) into the back of the upper buffer tank (Figure 17.3, #3). Loosen the upper clamp knobs of the rail assembly (Figure 17.3, #8 and #9) and slide the tank into place; tighten the knobs as before.

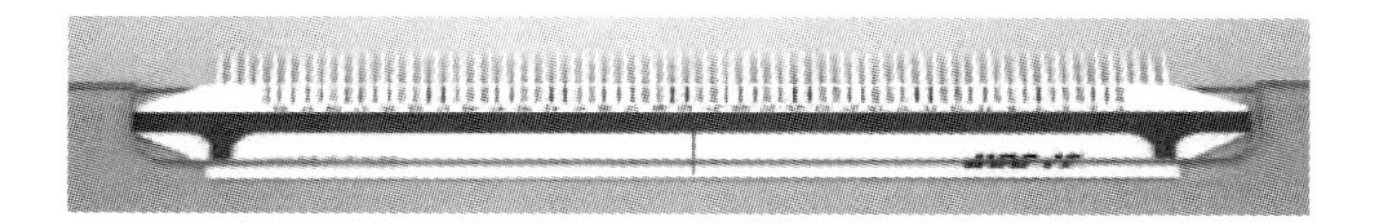

Figure 17.4
Sharkstooth comb. (Used with permission from LI-COR Biosciences, Lincoln, Nebraska.)

3. Open the door of the Model 4300 DNA analyzer and place the lower buffer tank (Figure 17.3, #11) at the base of the unit.
4. Place the gel apparatus on the DNA analyzer (against the heater plate) with the bottom of the gel cassette inside the lower buffer tank. The rail support pins (Figure 17.3, #13) will hold the gel cassette on the instrument.
5. Fill the upper and lower buffer tanks with 0.8× TBE running buffer prepared earlier (see "Gel Preparation" section, step 3). Before adding the running buffer, make sure the drain fitting in the upper buffer tank is closed.
6. Place the upper and lower buffer tank lids (Figure 17.3, #1 and 10) onto the tanks. Attach the high-voltage cable (Figure 17.3, #12) to the bottom of the upper buffer tank and insert the opposing end into the instrument chassis.

Starting the Run

Follow the manufacturer's manual and protocol for specific instructions on starting a new run using the e-Seq software. The e-Seq software automates almost the entire sequencing process by controlling the pre- and electrophoretic runs, and identifying the bases and their sequence along the mtDNA. After the pre-electrophoretic run, the e-Seq software will automatically pause the process for the user to load samples for analysis. The protocol outlined below highlights the major steps in the mtDNA sequencing analysis.

1. After the prerun, open the instrument door of the DNA analyzer and remove the upper buffer tank lid (Figure 17.3, #1).
2. Using a 20 cc syringe, flush the wells with buffer to remove any debris that may have settled during the pre-electrophoresis run.
3. Load the samples to be analyzed using a Hamilton syringe or a pipet with a flat 0.2 mm micropipet tip. Position the tip between the glass plates and slowly release the sample into the wells.
4. After loading the samples, replace the upper buffer tank lid, close the instrument door, and push the "start run" button.

Base Calling

For base calling and editing the data output, refer to the e-Seq user guide.

Results

In this exercise, you will be analyzing mtDNA data, specific to a case study, generated using infrared fluorescence detection (Table 17.4) or data that you have generated using mtDNA isolated from a human liver cell line or from buccal swabs. As the nucleotides are electrophoretically separated, the fluorescently labeled nucleotides (or bases) are excited by the laser light source, the emission captured by a detection system, and because of the dual-IR dye capability, the dual-dye automated sequencer permits simultaneous generation of two sequence ladders.

A Case Study

A very respected and dependable young man did not show up for work one day. When the young man did not show up for work on the second day, his employer called his home to see if there was something wrong. When no one answered the phone, the employer tried to contact the young man on his cell phone. The employer was only able to leave a voicemail message. The employer then called the young man's family only to find out that he had left for work yesterday morning at the usual time.

The family called the police, who, suspecting foul play, launched an investigation. By the end of the week the young man's body was found in an alley behind the building where he was employed. He had been beaten to death with a blunt-end object. The victim's body was sent to the medical examiner's office, where

TABLE 17.4
Mitochondrial DNA Typing Results

	Hypervariable Region 1 (HV1)													
Sample	16131	16185	16186	16191	16217	16225	16226	16280	16296	16313	16322	16362	16370	16392
Standard	T	C	C	C	T	C	A	C	C	C	A	T	G	T
DD201Q1	•	•	•	•	•	•	G	•	•	G	T	•	•	•
DD201Q2	A	•	•	•	•	•	G	•	•	G	•	•	•	•
DD201Q3	•	•	•	•	•	•	G	•	•	G	T	•	•	R
DD201K1	A	G	G	G	C	A	•	A	A	G	•	A	C	•
DD201K2	•	•	•	•	•	•	G	•	•	G	T	•	•	•

	Hypervariable Region 2 (HV2)												
	75	148	153	155	184	188	191	197	249	265	309.1	315.1	318
Standard	G	A	A	T	G	A	A	A	A	T	–	–	T
DD201Q1	C	G	•	•	•	•	•	G	•	C	G	G	•
DD201Q2	C	G	•	•	•	•	•	G	•	C	•	G	•
DD201Q3	C	G	•	•	•	•	•	G	•	C	•	G	•
DD201K1	C	•	T	G	A	T	G	•	T	C	G	G	A
DD201K2	C	G	•	•	•	•	•	G	•	C	G	G	•

Note: –, in the published reference sequence (standard, top row) there is no nucleotide at this position and the sample(s) has an insertion; •, the nucleotide is the same at this position as in the standard; R, both an A and G were observed at this position.

his clothing was removed and sent to the state's laboratory for DNA analysis. During the police investigation, a hammer was found in a dumpster in the alley where the body was found. The hammer was placed in a paper bag and sent to the laboratory for analysis. Investigators also determined that the young man was having a relationship with a married woman who also worked for the same employer. When the married woman was questioned, she told the police that her husband had recently learned of the affair and he had vowed to "straighten things out." The police spoke to the husband, who denied any knowledge of the young man's death. Considering all facts, the police arrested the husband and charged him with the murder of the young man. While in custody, the police collected a buccal swab from the husband and sent the sample to the laboratory for analysis. During the autopsy, the medical examiner collected a reference sample from the young man, which was also sent to the laboratory for analysis.

At the laboratory, several strands of hair were discovered on the victim's shirt (designated DD201Q1), on the handle grip of the hammer (designated DD201Q2), and a hair from the head of the hammer (designated DD201Q3). During microscopic examination of the hairs it was determined that the hair samples were candidates for mtDNA typing. The isolated mtDNA from the hair samples was purified, amplified by PCR, and a complete mtDNA profile obtained comprising the HV1 and HV2 regions. The mtDNA sequence for the evidentiary and known reference samples and the nucleotide substitutions with respect to the standard published reference sequence for each sample in this case are presented in Table 17.4.

Data Analysis

1. Each of the evidentiary hair samples was analyzed according to standard protocol prior to opening or handling of the known reference samples. The known samples were also analyzed according to standard protocol established by the laboratory.
2. All negative controls (i.e., reagent blanks, PCR blanks) remained free of contaminating DNA. All positive controls responded as expected.
3. For each sample, a complete mtDNA profile was obtained comprising nucleotide positions 15995 to 16400 (HV1) and 45 to 405 (HV2).
4. The mtDNA profiles of the three hairs (DD201Q1, DD201Q2, and DD201Q3) should be compared to the mtDNA profiles of the husband, the primary suspect (DD201K1), and the victim (DD201K2). The resulting mtDNA profiles are shown in Table 17.4.

Interpreting Test Results

Your group will work together to interpret the mtDNA profiles shown in Table 17.4. Compare the mtDNA typing results from the suspect's reference sample (DD201K1) to the mtDNA profiles generated from the hair samples. Also, compare the mtDNA results from the victim's known reference sample to the mtDNA typing results from the three hair samples. Each mtDNA profile from the known reference and evidentiary samples should be compared to the standard mtDNA sequence, also known as the Anderson sequence. After completing your analysis answer the questions at the end of this exercise.

Mitochondrial DNA Typing Results														
	Hypervariable Region 1 (HV1)													
Sample														
Standard														

Mitochondrial DNA Typing Results (continued)													
Sample	Hypervariable Region 2 (HV2)												
Standard													

Note: –, in the published reference sequence (standard, top row) there is no nucleotide at this position and the sample(s) has an insertion; •, the nucleotide is the same at this position as in the standard; R, both an A and G were observed at this position.

Questions

1. Why were the hair samples subjected to mtDNA typing when the laboratory in the case study above routinely performed short tandem repeat (STR) analysis?

2. Why were the evidentiary samples analyzed separately from the known referenced samples?

3. When the mtDNA profile from the suspect was searched against profiles in the mtDNA database, it was found that the sequence was observed once in the Caucasian database. Since the database includes over 4,800 sequences, what is the forensic significance of this frequency? What is the significance of a frequency reported as "not previously observed" in the current database?

4. When will reliable population frequency estimates for mtDNA types be available?

Chapter 18

Exercise 14

Assessment of DNA Typing Data

Introduction

There are two important steps in the assessment of DNA typing data: (1) visual examination of the gel images and (2) computer-assisted band size or allelic determination. Essentially, the gel image, whether generated by gel or capillary electrophoresis, is visually inspected to determine if the number, position, and intensity of the alleles for the allelic ladder, controls, and samples (i.e., known reference and evidence) are suitable for interpretation.

Visual Evaluation of the Gel Images

1. *Molecular weight and allelic ladders.* Visually examine the lanes of the gel image containing the molecular weight markers and the allelic ladders. The bands/peaks in these lanes must be of sufficient intensity to be used as molecular weight references for the positive allelic control (DNA from the K562 or GM9947A cell line; Fregeau et al., 1995), the known reference sample(s), and the questioned or evidentiary DNA bands/peaks. If portions of the molecular weight markers or the allelic ladders are not visible, the size or allelic designation of the evidentiary DNA bands/peaks cannot be determined in these areas of the gel image. If the banding patterns appear consistent, the amplified alleles are designated by noting which allelic ladder band(s)/peak(s) lines up with the sample band(s)/peak(s).
2. *Positive controls.* Visually examine the lane(s) of the gel image containing the positive control (i.e., DNA from the K562 or GM9947A cell line). There must be either one or two DNA bands/peaks for the DNA positive control, depending on which STR locus has been amplified. If the positive control does not exhibit the expected banding pattern or the number of bands or peaks for the locus under investigation, the typing data should not be assessed.
3. *Quality of banding patterns.* To assess the quality of the banding patterns, visually examine the lanes of the gel image containing the known or evidentiary DNA. DNA band irregularities in these lanes, such as increased band width (e.g., extremely broad bands) or "smiles" (e.g., pronounced band curvature), usually indicate potential mobility shifts during gel electrophoresis and will often compromise the interpretation of the data. For amplified products separated by capillary electrophoresis, the banding pattern observed in the electropherogram is examined not only for the presence of true alleles, but also for artifacts. The evaluating analyst must ensure that artifacts (i.e., spikes, "pull-ups," stutters, etc.) are not mislabeled as alleles. If the overall quality of the gel image is unsuitable for interpretation, no further comparisons are conducted. If the overall quality of the gel image is suitable for interpretation, a visual comparison is performed.
4. *Partial or incomplete profile.* If the evidence samples under comparison contain a partial profile (i.e., allele dropout at one or more loci) or an incomplete profile (i.e., locus dropout) due to degradation, inhibition, or limited DNA, the DNA profile may or may not be interpretable. However, all loci should be taken into account when making this determination using not only knowledge of the system, but also experience, if applicable.

5. *Mixtures.* An evidentiary sample often may contain DNA from more than one contributor or source. A sample that is determined to be consistent with a mixture may be determined by examining the number of alleles (bands/peaks) at each locus or, in the case of data generated by capillary electrophoresis, the peak height ratios at loci with more than one allele. A sample may also be considered to be a mixture of DNA from more than two individuals if three or more bands/peaks are observed at one or more loci or there is a distinct difference in peak heights at heterozygous loci. All loci should be taken into account when making this determination.
6. *Final assessment.* Based on the assessments of the quality and position of the DNA patterns from the gel image, decide which samples will be subjected to the computer-assisted band/peak sizing or allelic determination procedure.

Computer-Assisted Band/Peak Size Determination and Allelic Designation

The molecular weight determination or allelic designation of each DNA band/peak is carried out using Windows-based or MS DOS-based computer programs (e.g., GenoTyper, GeneScan Analysis, Gel-Pro, DNA Image Analysis) or by a fluorescent imaging system using specific software. The forensic DNA analyst is guided through the imaging and sizing procedures by text display on the computer screen. The computer software program enables an objective estimation of the sizes or allelic designation of the amplified DNA products in the positive control, the known reference samples, and each evidentiary sample. The sizing program ends by printing out the calculated sizes or molecular weight, in base pairs, and the allelic designation for each of the samples and for the allelic ladder. If the DNA bands/peaks for the positive control (amplified K562 or GM9947A DNA) in a particular profile (i.e., gel image or an electropherogram) are not within the acceptable size range, the profile in question should not be used for any conclusive match determinations.

Process of Sizing or Allelic Determination Using an Internal Standard

When analyzing the amplified products using capillary electrophoresis, a computer software program enables an objective estimation of the molecular weight or size, in base pairs, and the allelic designation of the amplified DNA products in the samples tested. For amplified products analyzed by gel electrophoresis, these determinations can be made manually by comparing the bands observed in the test samples (or crime scene samples) and the known reference samples to allelic ladders or the internal lane standards. The molecular weight or allele(s) determination of each band is performed as follows:

1. Using a ruler, measure the distance that each band (i.e., of the molecular weight standards, the known reference samples, and the test samples) has migrated from the bottom of the well (the origin) to the center of each DNA fragment.
2. To determine the molecular weight or size of each DNA band and the allelic designation, a standard curve is created using the distance traveled by each band (x-axis) and allelic size (y-axis) data from the molecular weight markers, the allelic ladder, and the internal standards. Record your data using the table provided at the end of this exercise. Then, using both the linear and semilog paper (provided at the end of this exercise), plot the distance that each DNA band has migrated (in mm) on the x-axis versus the size (in base pairs) on the y-axis for each band of the allelic ladder and internal standards. Record the molecular weight and the allelic designation of each DNA band in either table below. After each data point has been plotted on the graph paper, use a ruler and draw the best-fit line through the data points. Extend the line through the entire graph paper.
3. Compare the results of your graphs. Decide which graph should be used to estimate the size and allelic designation of the DNA bands observed from the test samples.
4. To estimate the size of a DNA band or allelic designation from the test sample, determine the distance the band migrated. Locate the distance migrated on the x-axis of your graph. Read "up" to the standard line, and then

follow the graph line over to the y-axis. This point is the approximate size (in base pairs) of the unknown DNA band/peak. Record the molecular weight of each DNA band or the allelic determination in either table below. Repeat this process for each DNA band.

5. Compare the sizes of the DNA bands of the test samples to those of the reference samples. Is there a match?

STR Typing Results

	Allelic Ladder			Test Samples (Crime Scene)			Reference Sample (Suspect)			Reference Sample (Victim)		
Band	Distance (mm)	Actual Size[a]	Allele	Distance (mm)	Actual Size[a]	Allele	Distance (mm)	Actual Size[a]	Allele	Distance (mm)	Actual Size[a]	Allele
1												
2												
3												
4												
5												
6												

	Internal Standard			Test Samples (Crime Scene)			Suspect 1			Suspect 2		
Band	Distance (mm)	Actual Size[a]	Allele	Distance (mm)	Actual Size[a]	Allele	Distance (mm)	Actual Size[a]	Allele	Distance (mm)	Actual Size[a]	Allele
1												
2												
3												
4												
5												
6												

[a] The actual size of the DNA bands is an approximation of the molecular weight in base pairs (bp).

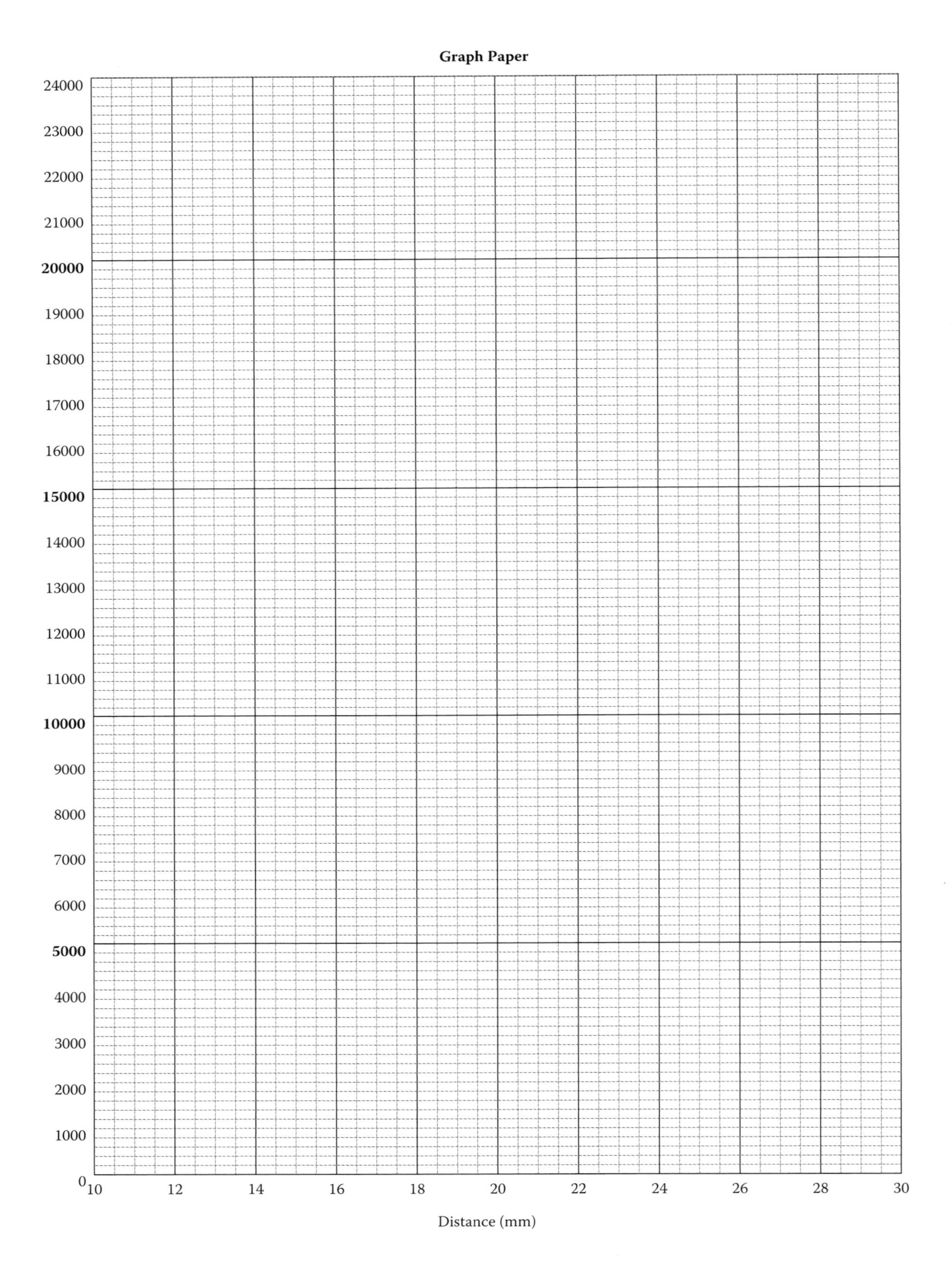
Graph Paper
24000
23000
22000
21000
20000
19000
18000
17000
16000
15000
14000
13000
12000
11000
10000
9000
8000
7000
6000
5000
4000
3000
2000
1000
0
Size (base pairs)
10
12
14
16
18
20
22
24
26
28
30
Distance (mm)

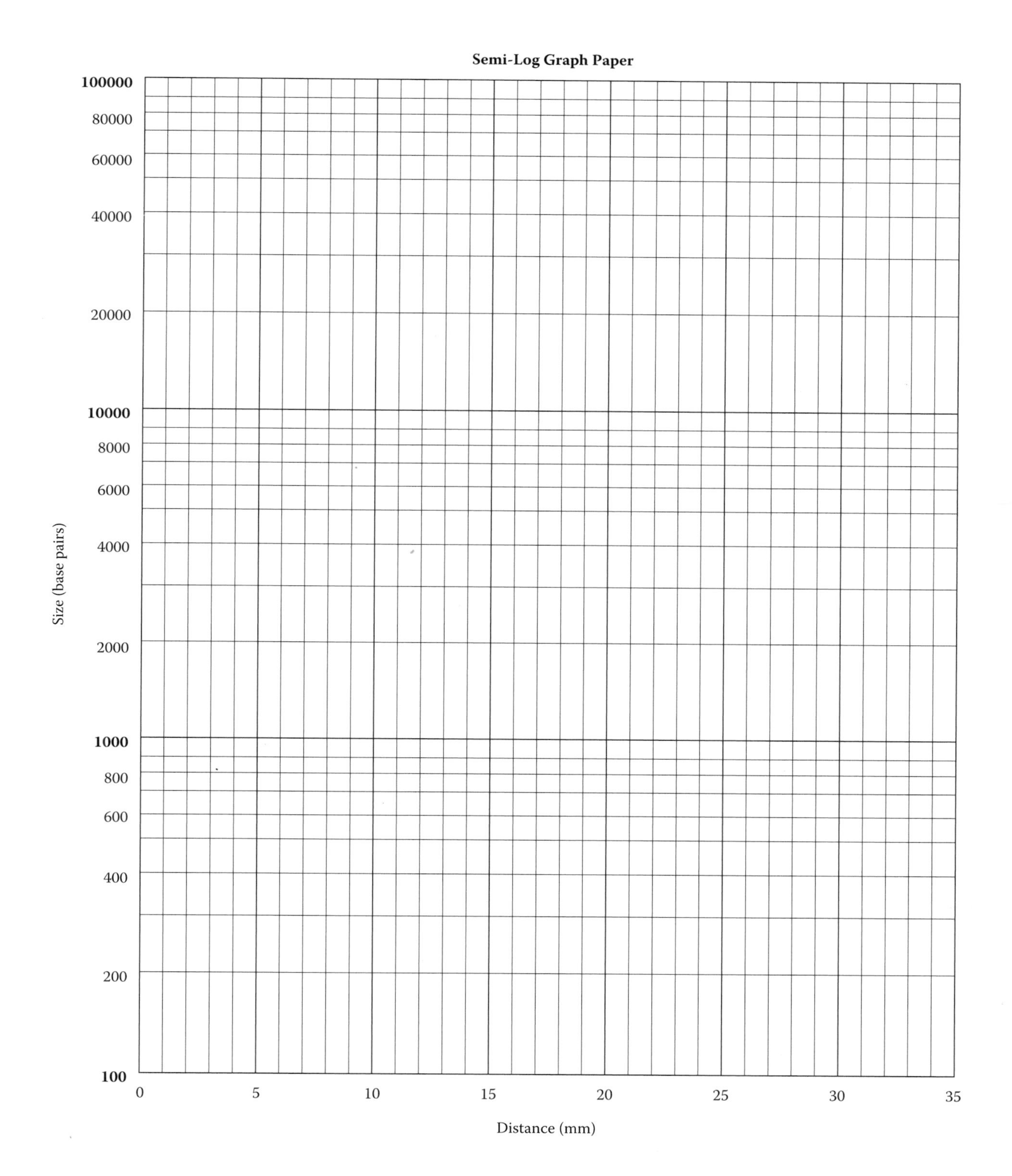
Semi-Log Graph Paper
100000
80000
60000
40000
20000
10000
8000
6000
4000
2000
1000
800
600
400
200
100
Size (base pairs)
0
5
10
15
20
25
30
35
Distance (mm)

Questions

1. What are the two important steps in the assessment of a DNA profile?

2. When the positive control fails to exhibit the expected number of DNA bands/peaks in a gel image, are the results of the other samples tested (i.e., known or evidentiary) interpretable? Reliable? Why or why not?

3. If the allelic ladder or molecular weight markers are not visible or the DNA bands/peaks in these lanes are very low in intensity, the DNA typing data cannot be assessed. Why?

Chapter 19

Single-Nucleotide Polymorphisms (SNPs) for Human Identification

An Emerging Forensic Tool

Genetic typing has revolutionized criminal investigations and has become a powerful tool in the identification of individuals in criminal and paternity cases. Several different genetic approaches and techniques have been developed over the past few years that are quite robust in generating a complete DNA profile, in spite of sample integrity. Current DNA typing techniques are able to reduce sample consumption and multiple manipulations, minimize resources and time to completion, and amplify and type up to 16 different loci simultaneously through multiplexing, while increasing the power of discrimination.

Currently, the state-of-the-art DNA typing technique is the polymerase chain reaction (PCR) amplification of short tandem repeats (STRs) (see Exercise 10). STR analysis is used internationally because of the small product size (i.e., allelic sizes range from 175 to 420 bp), the polymorphic nature of STRs, and because this typing process is amenable to semiautomation (Edwards et al., 1991; Budowle et al., 2001). Because the product length is shorter than those products produced previously using restriction fragment length polymorphism (RFLP) (see "An Overview of Forensic DNA Analysis," Chapter 1) analysis, evidentiary samples that are degraded are now typable using STR analysis.

However, there are still some forensically important evidentiary samples that are not amenable to STR analysis due to low concentrations of DNA, or the sample is too degraded to generate a reliable profile. Alternate technologies have been sought to overcome these inherent problems in typing difficult samples by focusing on the sequence of specific regions in the human mitochondrial genome. Mitochondrial DNA (mtDNA) analysis has been performed, but due to costs and labor involved, only a few laboratories have offered this service. In addition, the power of discrimination achieved through mtDNA typing is significantly lower than that obtained with STR analysis.

One promising class of genetic markers that may prove useful in forensic DNA analysis, particularly for typing degraded samples or other difficult-to-type samples containing low levels of DNA, is single-nucleotide polymorphisms (SNPs) (Kidd et al., 2006; Pakstis et al., 2007). SNPs are base substitutions, insertions, or deletions that occur at single positions within the human genome (Figure 19.1). SNPs are abundant in the human genome occurring, on average, about every 100 to 300 bases along a DNA strand. Their small product size (average range is 60 to 130 bp) is uniquely suitable for typing degraded or archaeological samples and can provide specific information about identity, lineage, ancestry, and phenotype. However, the potential use of SNPs in the forensic community has its limitations. SNPs are not currently represented in the Combined DNA Index System (CODIS) database, whereas STR loci are well established, and it would take a significant reduction in cost and enhanced abilities to resolve samples containing more than one contributor (i.e., mixtures) to replace the STR loci. Although SNPs are unlikely to replace STR analysis in the near future, SNP markers may serve an important role as an additional tool in the analysis of forensically challenging samples.

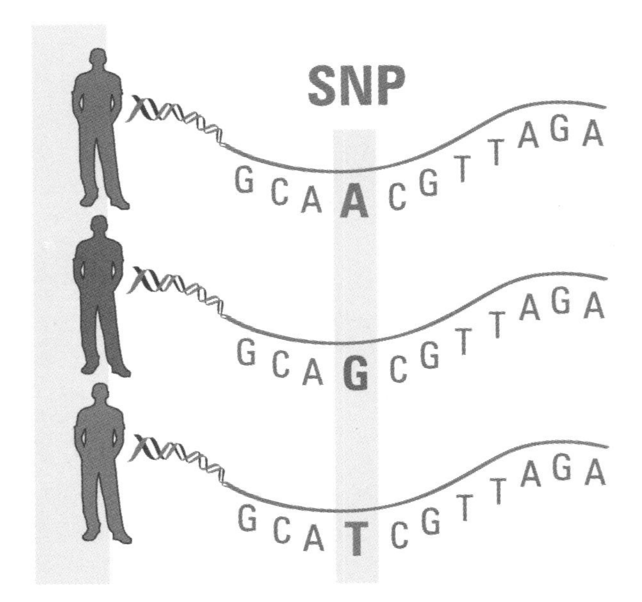

Figure 19.1
Variations of SNPs in the human population. SNPs are base substitutions, insertions, or deletions that occur at single positions within the human genome. SNPs, part of the genetic variation within a population, are abundant in the human genome occurring, on average, about every 100 to 300 bases along a DNA strand. (Courtesy of the Broad Institute, Inc.)

Several applications have been developed that examine the presence of SNPs by hybridization-based methods (i.e., hybridizing complementary DNA probes to the target SNP site) and enzyme-based methods (i.e., DNA ligase, DNA polymerase, and nucleases) (Rapley and Harbron, 2004; Olivier, 2005). One reliable SNP genotyping method that has recently been developed and is available commercially is the SNaPshot™ Multiplex System (Life Technologies, Grand Island, New York). In a single reaction tube, the SNaPshot kit examines up to 10 SNPs at specific sites along the DNA. The SNaPshot Multiplex System is a primer extension-based assay that uses a dideoxynucleotide (ddNTP) single-based extension of an unlabeled oligonucleotide primer(s). Each primer binds to its complementary strand or target DNA immediately upstream of the SNP nucleotide, and the DNA polymerase extends the primer by one fluorescently labeled ddNTP to the SNP nucleotide. Since the ddNTP is missing the 3′-hydroxyl end, no additional deoxynucleotides (dNTP) can be added. This incorporated nucleotide is detected by electrophoresis, using an ABI 3700 Genetic Analyzer (Applied Biosystems, Foster City, California), and ultimately allows the SNP allele to be determined using the GeneMapper software (Figure 19.2). And with each ddNTP labeled with a different fluorescent molecule, all four alleles can be detected in the same single reaction.

Novel SNPs have also been detected using the LI-COR infrared DNA analyzer (Model 4300) (see Exercise 13) in human mtDNA and in inbred rat strains (Punt et al., 2008; Smits et al., 2004). In the first study, Punt et al. selected two segments in the coding region (NADH dehydrogenase subunit 2 gene) of human mtDNA, as well as the hypervariable region 2 (HV2), and sequenced these regions by PCR using forward primers and LI-COR's IRDye™ labeled dideoxynucleotide terminators. Although no unique SNPs were found in HV2, two novel SNPs were identified in the dehydrogenase gene and appeared to be of Asian descent. In the other study, Smits et al. used enzymatic cleavage of heteroduplexes using the endonuclease CEL I for SNP discovery (Figure 19.3). Genomic DNA was isolated from cell samples using the Qiagen DNeasy Tissue Kit (see Exercise 6), transferred into 96-well plates, subjected to PCR, and the heteroduplexes cleaved with the endonuclease CEL I at the mismatched sites. The purified products were separated and sequenced on the 4300 DNA Genetic Analyzer.

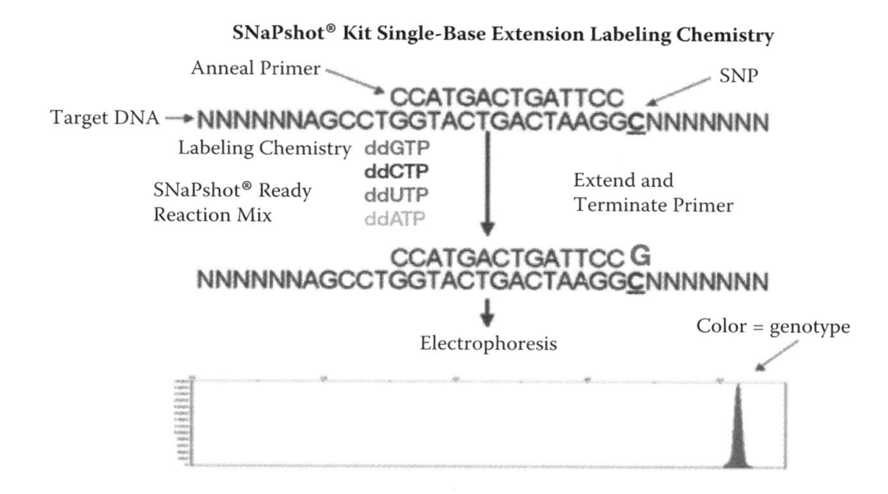

Figure 19.2
The SNaPshot Multiplex System: Single-base extension and termination. In the single-tube reaction, the primer binds upstream to the targeted SNP site and the DNA polymerase extends the primer by one nucleotide, adding a single ddNTP to its 3′ end. The fluorescence color readout determines which base was added. (Courtesy of Applied Biosystems.)

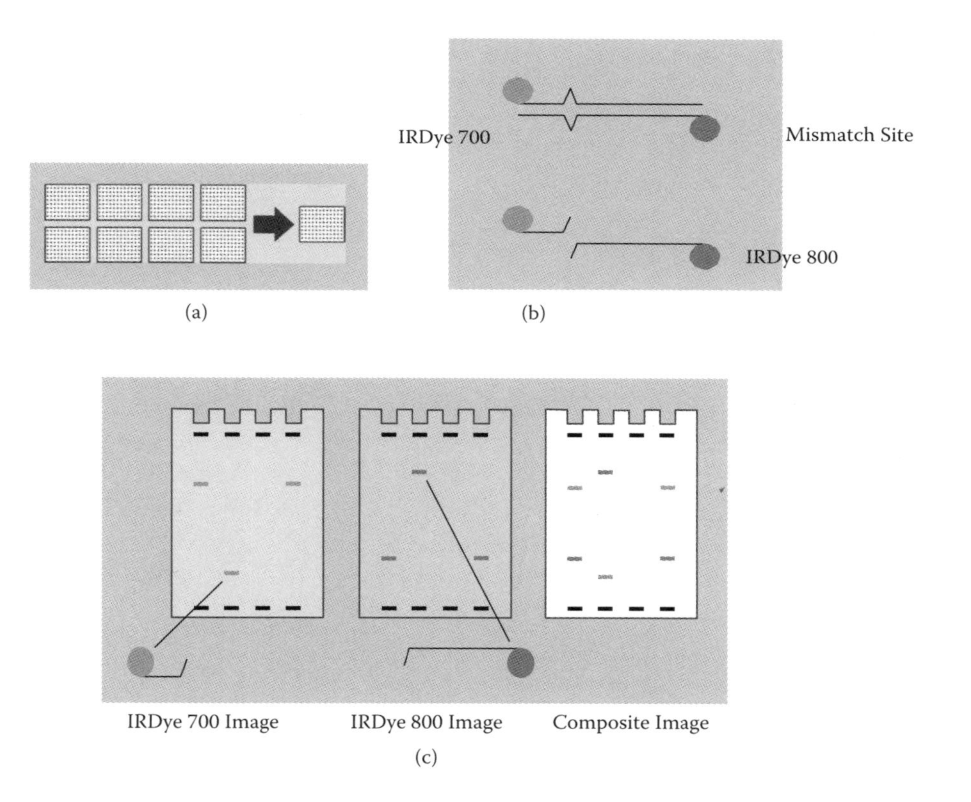

Figure 19.3
SNP discovery using the endonuclease CEL I. (a) DNA is extracted from both known reference and evidentiary samples and placed in microcentrifuge tubes or 96-well plates for high-throughput screening. (b) During PCR amplification two primers are labeled with a specific IRDye. After amplification the heteroduplexes are cut at the mismatch sites by CEL I. (c) The cut strands are separated by electrophoresis in the 4300 infrared DNA analyzer. SNPs in one fluorescence channel (IRDye 700) are confirmed by the presence of the opposite strand in the same lane in the other fluorescence channel (IRDye 800). Empty lanes indicate no genetic variation or SNPs in the sample analyzed. (Courtesy of LI-COR Biosciences.)

There are currently four different categories of SNPs that would benefit the forensic community: (1) identity testing or individual identification SNPs, (2) ancestry informative SNPs, (3) phenotype informative SNPs, and (4) lineage informative SNPs. The first group to be utilized in forensic casework is the individual identification SNPs. This group could be used to identify an individual that contributed to an evidentiary sample and used on difficult-to-type samples such as mixtures or samples that are degraded. However, these SNPs collectively yield very low probabilities of two individuals having the same multilocus genotype; thus, the question is how many SNPs would be needed to match the power of discrimination linked to STR analysis. One laboratory has reported that a panel of 40 SNPs could yield an average match probability in most populations of less than 1×10^{-16} or greater than 1 in 1 quadrillion (Pakstis et al., 2007).

The lineage informative SNPs are thought to be the next category to be utilized in the forensic community. This group of tightly linked SNPs could serve to identify relatives with higher probabilities than with simple biallelic SNPs. The ancestry informative SNPs and the phenotype informative SNPs are thought to be utilized in specific applications that would provide useful investigative leads. The ancestry informative SNPS would give a high probability of an individual's ancestry being from one part of the world or being derived from two or more areas of the world. The phenotypic informative SNPs provide a high probability of predicting phenotype, in some instances higher than the ancestry informative SNPs, such as skin color, hair color, and eye color. Surely, more research will lead to the identification and development of SNPs that could enhance forensic leads and investigations.

The purpose of this section was to provide an overview of SNPs, the various types of SNPs, the methodology used to discover and identify SNPs, and their potential use in the forensic community. Although a detailed experimental protocol and comprehensive list of materials were not provided, it is hoped that this overview will encourage readers to refer to the cited articles and determine if their laboratory, either at an academic institute or a state, federal, or privately funded laboratory, has the resources and capabilities to perform this sort of analysis.

References and Suggested Readings

Allery, J.P., N. Telmon, R. Mieusset, A. Blanc, and D. Rougé. 2001. Cytological detection of spermatozoa comparison of three staining methods. *J. Forensic Sci.* 46(2):349–351.

Anderson, S., A.T. Bankier, B.G. Barrell, M.H. de Bruijn, A.R. Coulson, J. Drouin, I.C. Eperon, D.P. Nierlich, B.A. Roe, F. Sanger, P.H. Schreier, A.J. Smith, R. Staden, and I.G. Young. 1981. Sequence and organization of the human mitochondrial genome. *Nature* 290:457–465.

Andrews, R.M., I. Kubacka, P.F. Chinnery, R.N. Lightowlers, D.M. Turnbull, and N. Howell. 1999. Reanalysis and revision of the Cambridge reference sequence for human mitochondrial DNA [letter]. *Nat. Genet.* 23:147.

Auvdel, M.J. 1986. Amylase levels in semen and saliva stains. *J. Forensic Sci.* 31(2):426–431.

Budowle, B., B. Shea, S. Niezgoda, and R. Chakraborty. 2001. CODIS STR loci data from 41 sample populations. *J. Forensic Sci.* 46:453–489.

Butler, J.M., and B.C. Levin. 1998. Forensic applications of mitochondrial DNA. *Trends Biotechnol.* 16:158–162.

Edwards, A.A., H.A. Civitello, H.A. Hammond, and C.T. Caskey. 1991. DNA typing and genetic mapping with trimeric and tetrameric tandem repeats. *Am. J. Hum. Genet.* 49:746–756.

Edwards, A., H.A. Hammond, L. Jin, C.T. Caskey, and R. Chakraborty. 1992. Genetic variation at five trimeric and tetrameric tandem repeat loci in four human population groups. *Genomics* 12:242–253.

Fregeau, C.J., R.A. Aubin, J.C. Elliott, S.S. Gill, and R.M. Fourney. 1995. Characterization of human lymphoid cell lines GM9947 and GM9948 as intra- and interlaboratory reference standards for DNA typing. *Genomics* 28:184–197.

Gaensslen, R.E. 1983. *Sourcebook in forensic serology, immunology, and biochemistry*. U.S. Government Printing Office, Washington, DC.

Gilbert, D.A., Y.A. Reid, M.H. Gail, D. Pee, C. White, R.J. Hay, and S.J. O'Brien. 1990. Application of DNA fingerprint for cell-line individualization. *Am. J. Hum. Genet.* 47:499–514.

Gill, P., A.J. Jeffreys, and D.J. Werrett. 1985. Forensic application of DNA 'fingerprints.' *Nature* 318:577–579.

Jaffe, M. 1886. Ueber den Niederschlag, Welchen Pikrirtsure in Normalen Ham Erzeugt and Ber eirie Neue Reaction des Kreatinins. *Z. Physiol. Chem.* 10:391.

Jeffreys, A.J., V. Wilson, R. Neumann, and J. Keyte. 1988. Amplification of human minisatellites by polymerase chain reaction: towards DNA fingerprint of single cells. *Nucleic Acid Res.* 16:10953–10971.

Jeffreys, A.J., V. Wilson, and S.L. Thein. 1985. Individual-specific fingerprint of human DNA. *Nature* 316:76–79.

Kasai, K., Y. Nakamura, and R. White. 1990. Amplification of a variable number of tandem repeat (VNTR) locus (pMCT118) by the polymerase chain reaction (PCR) and its application to forensic science. *J. Forensic Sci.* 35:1196–1200.

Kennedy, S. 2009. Isolation of DNA from FFPE samples without paraffin removal. *BioTechniques* 21.

Kidd, K.K., A.J. Pakstis, W.C. Speed, E.L. Grigorenko, S.L.B. Kajuna, N.J. Karoma, S. Kungulilo, J. Kim, R. Lu, A.Odunsi, F. Okonofua, J. Parnas, L.O. Schulz, O.V. Zhukova, and J.R. Kidd. 2006. Developing a SNP panel for forensic identification of individuals. *Forensic Sci. Int.* 164:20–32.

Landsteiner, K. 1962. *The specificity of serological reactions*. Dover Publications, New York.

Lathia, D., and M. Brendeback. 1978. Influence of thiocyanate ions on starch-iodine reaction used for the estimation of a-amylase activity. *Clin. Chim. Acta* 82:209–214.

Lennard, C., and M. Stoilovic. 2004. Application of forensic light sources at the crime scene. In *The practice of crime scene investigation*, ed. J. Horswell. CRC Press, Boca Raton, FL, pp. 97–123.

Lozzio B.B., and C.B. Lozzio. 1979. Properties and usefulness of the original K-562 human myelogenous leukemia cell line. *Leuk. Res.* 3:363–370.

Miller, S.A., D.D. Dykes, and H.F.D. Polesky. 1988. A simple salting out procedure for extracting DNAs from human nucleated cells. *Nucleic Acid Res.* 16:1215.

Nakamura, Y., M. Carlston, V. Krapcho, and R. White. 1988. Isolation and mapping of a polymorphic DNA sequence (pMCT118) on chromosome 1p (D1S80). *Nucleic Acid Res.* 16:9364.

Nakamura, Y., S. Gillilan, P. O'Connell, M. Leppert, G.M. Lathrop, J-M. Lalouel, and R. White. 1987b. Isolating and mapping of a polymorphic DNA sequence PYNH24 on chromosome 2 (D2S44). *Nucleic Acid. Res.* 15:10073.

Nakamura, Y., M. Leppert, P. O'Connell, R. Wolfe, T. Holm, M. Culver, C. Martin, E. Fujimoto, M. Hoff, E. Kumlin, and R. White. 1987a. Variable number of tandem repeat (VNTR) markers for human gene mapping. *Science* 235:1616–1622.

Olivier, M. 2005. The invader assay for SNP genotyping. *Research* 573(1–2):103–110.

Pakstis, A.J., W.C. Speed, J.R. Kidd, and K.K. Kidd. 2007. Candidate SNPs for a universal individual identification panel. *Hum. Genet.* 121:305–317.

Punt, C., E. Smalley, and T. Salerno. 2008. Identification of single nucleotide polymorphisms in the coding region of human mitochondrial DNA. In *Journal of Undergraduate Research*. Undergraduate Research Center, Minnesota State University, Mankato.

Rapley, R., and S. Harbron. 2004. *Molecular analysis and genome discovery*. John Wiley & Sons, Chichester, UK.

Sanger, F.S., S. Nilken, and A.R. Coulson. 1977. DNA sequencing with chain-terminating inhibitors. *Proc. Natl. Acad. Sci. USA* 74:5463–5467.

Schill, W.B., and G.F.B. Schumacher. 1972. Radial diffusion in gel for micro determination of enzymes. I. Muranidase, alpha-amylase, DNase 1, RNase A, acid phosphatase, and alkaline phosphatase. *Anal. Biochem.* 46 (2):502–533.

Smits, B.M.G., B.F.M. van Zutphen, and R.H.A. Plasterk. 2004. Genetic variation in coding regions between and within commonly used inbred rat strains. *Genome Res.* 1285–1290.

Southern, E. 1975. Detection of specific sequences among DNA fragments separated by gel electrophoresis. *J. Mol. Biol.* 98:503.

Stoilovic, M. 1991. Detection of semen and blood stains using polilight as a light source. *Forensic Sci. Int.* 51:289–296.

Vandenburg, N., and R.A.H. Oorschot. 2006. The use of polilight in the detection of seminal fluid, saliva, and bloodstains and comparison with conventional chemical-based screening tests. *J. Forensic Sci.* 51:361–370.

Virkler, K., and I.K. Lednev. 2009. Analysis of body fluids for forensic purposes: from laboratory testing to non-destructive rapid confirmatory identification at a crime scene. *Forensic Sci. Int.* 188:1–17.

Wahls, W.P., L.J. Wallace, and P.D. Moore. 1990. Hypervariable minisatellite DNA is hotspot for homologous recombination in human cells. *Cell* 60:95–103.

Warne, D., C. Watkins, P. Bodfish, K. Nyberg, and N.K. Spurr. 1991. Tetranucleotide repeat polymorphism at the human β-actin related pseudogene 2 (actbp2) detected using the polymerase chain reaction. *Nucleic Acids Res.* 19:6980.

Appendix

Composition of Buffers and Solutions

Ammonium persulfate solution (10%): Add 0.1 g of ammonium persulfate to 1 ml of deionized water. Prepare fresh before use.

Analytical gel visual markers: KpnI-digested adenovirus II DNA. Fragment lengths (in base pairs): 1,086, 1,699, 2,049, 2,339, 3,648, 5,167, 5,758, 6,478, and 7,713.

AP spot test solution: If purchased from a commercial supplier, follow product insert regarding preparation of solution and storage requirements. Add 0.26 g of the powder to 10 ml of sterile distilled water. Label test tube or dropper bottle containing the mixed solution with the date and user/group's name. Store the solution in an amber bottle or protect the container from light by covering with foil. Use within 24 h of preparation.

If the solution is being manually prepared, use the following directions. Prepare buffer by adding 5 ml of glacial acetic acid to 500 ml of distilled water. Add 10 g of sodium acetate (anhydrous; 0.24 M).

Solution A: Add 0.63 g of sodium α-naphthyl phosphate (0.25% w/v) to 250 ml of buffer.

Solution B: Add 1.25 g of naphthanil diazo blue B (0.5% w/v) to 250 ml of buffer.

Both solutions are combined to make the final working solution in the spot test (or used separately). When solutions A and B are combined, the reagent is not as stable as when stored separately.

Blotting pads (chemiluminescent detection): 11 × 16 cm blotting pads.

Chelex: Weigh out 1 g of Chelex 100 (100–200 mesh, sodium form). Add 50 mM Tris to the 1 g of Chelex to make 10 ml of solution. Adjust the pH to 11 using 4 N NaOH. Store at room temperature (expiration: 3 months).

Dithiothreitol (DTT) 1 M: Dissolve 1.54 g of DTT in 100 μl of sterile sodium acetate, pH 5.2, and bring final volume to 10 ml with sterile deionized water. Store in 100 μl aliquots in 1.5 ml microcentrifuge tubes at –20°C (expiration: 1 year).

Ethidium bromide (5 mg/ml): *Warning*: Mutagenic substance. Dissolve 1.0 g of ethidium bromide in 200 ml of distilled water. Keep bottle wrapped in foil to protect contents from light.

Ethylenediaminetetraacetic acid (EDTA) 0.5 M: Add 80 ml of water to 18.62 g of disodium EDTA-$2H_2O$. Slowly add NaOH pellets to lower the pH to 8.0. When fully dissolved, add more NaOH to bring the pH to 8.0. Adjust final volume to 100 ml. Autoclave and store at room temperature (expiration: 6 months).

Extraction buffer (amylase assay): Dissolve 8.42 g of NaCl and 2.38 g of HEPES in 900 ml of deionized water. Use NaOH to bring the pH to 7.2. Adjust final volume to 1.0 l with deionized water and store at 4°C.

Gel buffer (amylase assay): Dissolve 5.4 g of NaH_2PO_4, 3.9 g of $NaHPO_4$, and 0.4 g of NaCl in 900 ml of deionized water. Use NaOH to adjust the pH to 6.9 and bring to a final volume of 1.0 l with deionized water.

HaeIII: Restriction endonuclease used during RFLP analysis. HaeIII recognizes the four-base sequence GGCC cleaving between the G and C to create a blunt-ended DNA strand.

Hydrogen peroxide (3%): Dilute 10 ml of 30% hydrogen peroxide to 100 ml with deionized water. Store at 4°C (expiration: 3 months or when the positive control fails to give a positive result).

Iodine solution (amylase assay): Using mild heat, dissolve 6.6 g of potassium iodide (KI) and 10.16 g of iodine (I_2) in 120 ml of deionized water. Store at room temperature. For use, dilute 0.2 ml with 50 ml of deionized water.

K562 DNA (HaeIII-digested): HaeIII-digested K562 DNA is available commercially at 25 ng/μl; used as cell line control for post-restriction digestion and analytical gels.

K562 DNA standard (uncut): The typing grade K562 DNA is available commercially and is diluted to 20 ng/μl prior to use.

Kernechtrot staining solution (KS): Dissolve 5.0 g of aluminum sulfate in 100 ml of hot distilled water. Immediately add 0.1 g of nuclear fast red and stir with a glass rod. Allow the solution to cool and filter. The solution is stable at room temperature for up to 6 months; may need to be refiltered after standing for a period of time.

Lambda DNA (HindIII-digested): Commercially available HindIII-digested DNA that is used as a visual marker for yield and post-restriction endonuclease digestion gels. Fragment lengths in base pairs are 125, 564, 2,027, 2,322, 4,361, 6,557, 9,416, and 23,130.

Lambda DNA (uncut): Commercially available DNA for preparation of yield gel quantitation standards. Length in base pairs is 48,502.

Leucomalachite green (LCG) stock solution: Add 0.25 g of leucomalachite green and 5 g of zinc dust to 150 ml of distilled water followed by the addition of 100 ml of glacial acetic acid. Mix, add a few boiling chips, and boil under reflux for 2 to 3 h or until the solution has lost all of its color. Decant the cooled solution into a bottle containing some zinc to keep the solution in the reduced form.

Loading buffer: 50% glycerol (v/v)/0.1% (w/v) bromophenol blue/0.1 M EDTA. Prepare 50 ml of 100% glycerol, 0.1 g of bromophenol blue, and 20 ml of 0.5 M EDTA and bring to a final volume of 100 ml with TE.

Loading buffer CH: 10 mM Tris-Cl, pH 7.5/10% glycerol (v/v)/0.02% (w/v) bromophenol blue/20 mM EDTA. Prepare 400 μl of 2 M Tris-Cl, pH 7.5, 5 ml of 100% glycerol, 10 mg of bromophenol blue, and 2 ml of 0.5 M EDTA and bring to a final volume of 50 ml with distilled water. *Note*: Loading buffer CH is used only for dilution of molecular weight markers for chemiluminescent detection.

Luminol solution: Dissolve 5.0 g of sodium carbonate, 0.1 g of 5-amino-2,3-dihydro-1,4-phthalazinedione (luminol), and 10 ml of 3% hydrogen peroxide in 90 ml of deionized water. Dispense into an atomizer. Make just prior to use. Luminol is also available in tablet form from vendors, which is easily dissolved in water prior to use.

Molecular weight markers (chemiluminescent detection): DNA sizing markers that are available commercially; containing a tube with the molecular weight markers in solution and a tube of loading buffer.

Phenol/chloroform/isoamyl alcohol (100/100/4): Melt 100 g of phenol at 65°C and pour into a Bellco glass bottle. Add 200 mg 8-hydroxy-quinoline and mix the solution thoroughly. Add an equal volume of 1.0 M Tris, pH 7.5, transfer to a separatory funnel, mix, and let the phases separate. Drain the lower phenol layer into the Bellco bottle. Drain the upper aqueous phase into a waste beaker. Add an equal volume of 0.01 M Tris, pH 7.5, to the phenol, transfer to the separatory funnel, and mix. Capture the lower phase in the bottle. Capture the upper phase and determine its pH. If the upper phase pH is 7.5, cease equilibration procedures. If the pH is less than 7.5, repeat the extraction(s) with 0.01 M Tris until the pH of the upper phase is 7.5. Combine the equilibrated phenol with a solution composed of 100 ml of chloroform and 4 ml of isoamyl alcohol. Cover the solution with 0.01 M Tris and store at 4°C. This solution (in the ratio of 25:24:1) is also available commercially from various sources.

Phenolphthalein stock solution: In a 500 ml round-bottom flask place 2.0 g of phenolphthalein, 20 g of potassium hydroxide, 20 g of zinc granules, and 100 ml of deionized water. Place the flask on a heating mantle and reflux the solution (a technique involving the condensation of vapors and the return of this condensate to the system from which it originated) for 2 to 4 h, or until the solution is colorless. Store in an amber dropper bottle containing a small amount of zinc at 4°C (expiration: 6 months).

Phosphate-buffered saline (PBS): An isotonic salt solution frequently used to wash residual growth medium from a cell culture monolayer. 5× PBS (per liter) = 40 g of NaCl, 1.0 g of KCl, 5.75 g of Na_2HPO_4, 19 g of KH_2PO_4; autoclave; dilute aseptically to 1× with sterile H_2O prior to use.

Picroindigocarmine solution (PICS): Dissolve 1.0 g of indigocarmine dye in 300 ml in a saturated solution of commercially available picric acid. Filter and store for up to 6 months at room temperature.

Proteinase K (20 mg/ml): Dissolve 500 mg of proteinase K in a small volume of distilled water and bring to a final volume of 25 ml with distilled water. Dispense into convenient sized aliquots and freeze.

Sarkosyl (20%): Dissolve 20 g of N-lauroylsarcosine in distilled water and bring to a final volume of 100 ml. Sterilize by filtration using a sterile 0.45 μm filter.

Sodium acetate (2 M): Dissolve 41.02 g of sodium acetate in distilled water and bring to 200 ml with distilled water. Adjust the pH to 7.0 with concentrated HCl and adjust final volume to 250 ml with distilled water. Autoclave and store at room temperature.

Sodium dodecyl sulfate (SDS) (20% (w/v)): Add 200 g of sodium dodecyl sulfate to 700 ml of water and heat to 65°C to dissolve. Bring to a final volume of 1.0 l with distilled water.

Sodium hydroxide (5%): In an exhaust hood, dissolve 0.5 g of sodium hydroxide pellets in 10 ml of deionized water. This is an exothermic reaction—caution is needed.

Stain extraction buffer: 10 mM Tris-Cl/0.1 M NaCl/2% SDS/10 mM EDTA/39 mM DTT. Dissolve 1.21 g of Tris and 5.84 g of NaCl in 500 ml of distilled water. Add 100 ml of 20% SDS, 20 ml of 0.5 M Na_2EDTA-$2H_2O$, and adjust pH to 8.0 with HCl. Bring to a final volume of 1.0 l with distilled water. Supplement with DTT before use. To 100 ml of the above solution, add 0.6 g of powdered DTT and stir until dissolved. Store at room temperature. The final solution is good for no more than 2 weeks.

1× TAE: 40 mM Tris-acetate, pH 8.3/1 mM EDTA. Prepare 50 ml of 2OX TAE and 950 ml of distilled water.

1× TBE buffer: Disolve 54.0 g of Tris base and 27.5 g of boric acid in 800 ml of distilled water. Add 20 ml of 0.5M EDTA (pH 8.0). Bring the final volume to 1.0 liter (l) with distilled water and filter sterlize through 0.22 μm membrance. The final solution is a 5× stock solution and can be diluted to 1× prior to use.

TE buffer: 10 mM Tris-Cl, pH 7.5/0.1 mM EDTA. Prepare 1.210 g of Tris base and 0.037 g of Na_2EDTA. Dissolve Tris in 800 ml of distilled water and adjust the pH to 7.5 with HCl. Add EDTA, check the pH, and adjust to 7.5 if required. Bring the final volume to 1.0 l with distilled water and autoclave.

TE-9 buffer: 0.5 M Tris, pH 9.0/20 mM EDTA/10 mM NaCl.

TNE (pH 8.0): 10 mM Tris-Cl, pH 7.5/0.1 M NaCl/1 mM EDTA. Prepare 2.5 ml of 2 M Tris-Cl, pH 7.5, 10 ml of 5 M NaCl, and 1 ml of 0.5 M EDTA, pH 8.0. Add distilled water to 400 ml. Titrate to pH 8.0 with 0.1 N NaOH. Bring to a final volume of 500 ml with distilled water and autoclave.

Tris (2 M): Dissolve 242.2 g of Tris base in 800 ml of distilled water. Adjust to pH 7.5 with concentrated HCl. Bring the final volume to 1.0 l with distilled water and autoclave.

Appendix

Using a Micropipet

Introduction

Most chemical reactions in forensic DNA analysis are performed in small volumes of liquid, partly because DNA is available only in small quantities and because reagents and enzymes are expensive. The various reactions are performed in small microcentrifuge tubes (0.2–1.5 ml) in volumes as small as 0.5 µl. Consequently, the forensic DNA analyst must be able to dispense such small volumes correctly and accurately. Dispensing such volumes of liquids is accomplished using micropipets. The series of steps below provides an overview of the structure and use of the micropipet.

Structure of the Micropipet

Take a micropipet in your hand and become familiar with the various parts (see Figure B.1). On top of the micropipet is the plunger button for withdrawing and dispensing liquids. The second button, the tip ejector button, allows the user to eject the disposable tip, thus eliminating the need to touch the tip or liquid. An inset wheel or knob, the volume adjustment dial, permits the user to adjust the volume, whereas the dial with numbers (digital volume indicator) indicates the volume that has been selected. The plastic shaft at the base of the micropipet holds the disposable tip.

The micropipets are available in different ranges or capacities:

- P20—up to 20 µl (up to 0.02 ml)
- P200—20–200 µl (up to 0.2 ml)
- P1000—200–1,000 µl (up to 1.0 ml)

Micropipets are expensive and can be easily damaged if not handled properly. When using the micropipet always adhere to the follow rules:

- Never rotate the volume adjustment knob below the lower limit or above the upper limit.
- Never lay the micropipet on the bench; always replace in the stand when not in use. This prevents liquid from entering the pipet and causing damage.
- Never immerse the plastic shaft of the pipet into fluid without a tip in place. Always use a new tip for each different reagent. Use the proper size tip for each pipettor.
- Always slowly release the plunger when withdrawing or dispensing liquids.

Before using the pipet, familiarize yourself with the feel of the pipet. Hold the pipet in your writing hand. With your thumb, slowly lower and raise the plunger. As you press down on the plunger you will feel resistance at the "first stop," but you can continue to press until the plunger stops. The pipet is filled by

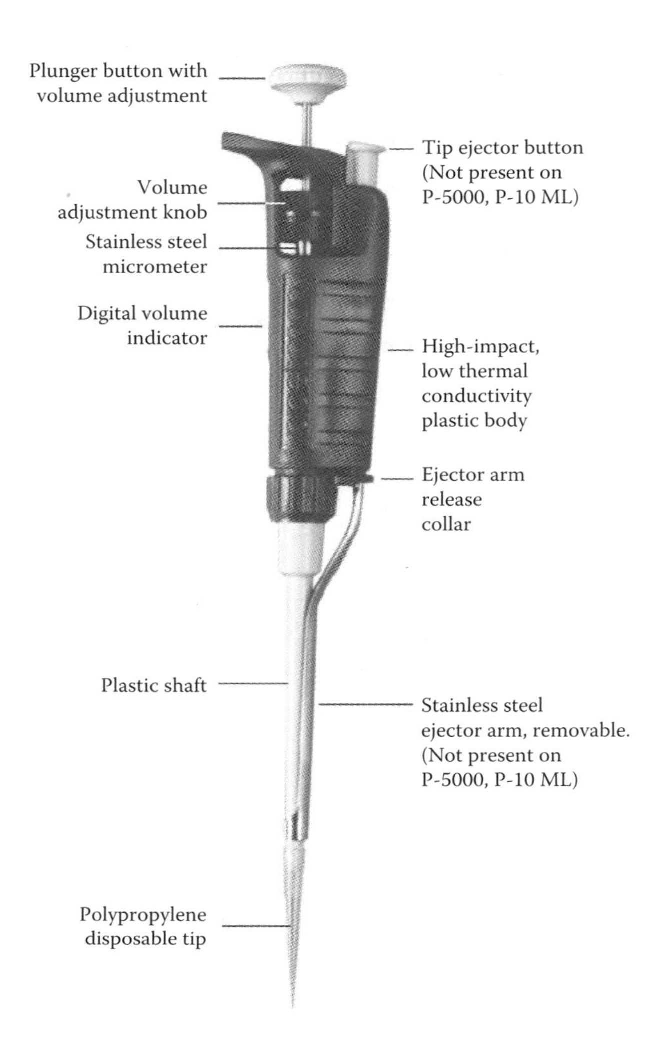

Figure B.1
A Rainin classic pipetman. (From Rainin Instrument, LLC. With permission.)

pressing the plunger down to the first stop and slowly releasing the plunger. The first stop plus the second stop or end will empty the pipet.

Using the Pipet

1. Check the top of the micropipet's plunger button to select the pipet that you will need. Use a pipet with a volume greater than the amount to be pipetted. Refer to the range of sizes above.
2. To select the volume of liquid to be pipetted, rotate the volume adjustment knob until the digital indicator reaches the desired volume.
3. Place a disposable tip on the plastic shaft of the micropipet. Press firmly to ensure that the tip is in place.
4. Press the plunger down to the first stop. Hold the micropipet vertically and place the disposable tip into the liquid to be pipetted. It is only necessary to place the tip in the liquid to a depth of several millimeters.
5. Slowly release the plunger button to its original position. Make sure the liquid is drawn into the tip.
6. Withdraw the tip from the liquid.

7. To dispense the liquid sample, place the tip against the wall of the receiving tube and press the plunger down to the first stop, then to the final stop to dispense any remaining liquid.
8. While the plunger is still pushed sown, remove the tip from the tube and allow the plunger to slowly return to its original position.
9. Discard the disposable tip into a waste container by pressing the tip ejector button.

Appendix C

Glossary

Allele: An alternative form of a gene occupying a given location on a chromosome that determines alternative characteristics in inheritance.

Allele-specific oligonucleotide (ASO): A short, specific DNA sequence that is used as a probe in the AmpliType PM/DQA1 test to detect a unique sequence.

Allelic dropout: When an STR (*see* short tandem repeat) test detects only one of the two alleles from a particular contributor at a given locus; occurs when the quantity of the DNA is relatively low or degraded. Allelic dropout complicates the process of interpretation because the analyst must decide whether the mismatch between the two samples is due to genetic differences or simply the failure of the test to detect all of the alleles in the sample.

Alternate light source (ALS): A device, used for the detection of bodily fluids, which consists of a light source (such as a laser or incandescent bulb) and a filter or combination of filters that enable all but the selected wavelengths of light to be screened out that delivers the light to the evidentiary sample being examined. Without the use of appropriate viewing accessories (such as protective goggles, if the wavelengths of light being used are potentially harmful, or goggles equipped with a filter to screen out the harmful wavelengths), urine, like some other bodily fluids, will absorb the light and fluoresce in the dark.

Alu element: A short stretch of DNA (approximately 300 bp in length) originally characterized by the action of the Alu restriction endonuclease. Alu elements are the most abundant family of repeats (or repetitive sequences) in the human genome, with over 1 million copies comprising 10% of the genome.

Amino acids: Building blocks of proteins. Each protein consists of a specific sequence of amino acids. There are 20 common amino acid molecules that can make up proteins.

AmpliTaq: Recombinant form of the naturally-occurring thermostable DNA polymerase from the organism *Thermus aquaticus*.

Analytical threshold: The minimum height requirement at and above which detected STR peaks can be reliably distinguished from background noise. STR peaks above this threshold are considered true alleles or possible artifacts.

Anneal: The base pairing of complementary polynucleotides to form a double-stranded molecule.

Antiparallel: The manner in which two complementary polynucleotides base pair to one another; the 5′ and 3′ ends of each molecule are reversed in relation to each other, so that the 5′ end of one strand is aligned with the 3′ end of the other strand. Antiparallel base pairing accompanies the formation of double-stranded DNA (DNA:DNA), double-stranded RNA (RNA:RNA), and DNA-RNA hybrids (DNA:RNA).

Artifacts: A non-allelic product of the PCR amplification process (e.g., stutter, non-template nucleotide addition, or other nonspecific product), an anomaly of the detection process (e.g., pull-up), or a by-product of primer synthesis (e.g., dye blob).

Autoradiograph (also autoradiogram): A photographic record of the spatial distribution of radiation in an object or specimen. The autoradiograph is made by placing the object (e.g., a Southern blot hybridized with a DNA probe) very close to a photographic film or emulsion.

Autoradiography: The process by which an autoradiograph is made.

Avidin (streptavidin) enzyme conjugate (HRP-SA): A non-isotopic detection system used with the AmpliType PM/DQA1 typing system. Biotin is covalently attached to each primer pair. The biotin-labeled amplified products are allowed to hybridize to the DNA probes immobilized on the nylon test strips. The strips are reacted with the enzyme horseradish peroxidase (HRP) covalently bound to streptavidin (SA). This HRP-SA conjugate can only bind to hybridized or double-stranded targets. If hybridization occurs, the HRP-SA conjugate will react with the colorless substrate, causing a blue color to develop. The spot remains colorless if no hybridization occurred—hence a negative response.

Bacteriophage lambda: A virus that infects and is propagated in a bacterial host, often used for cloning purposes. Among the best characterized and widely exploited are derivatives of the A bacteriophage.

Base: One of five molecules that make up the informational content of DNA and RNA. In DNA, bases pair across the two chains of the double helix: adenine (A) with thymine (T), and guanine (G) with cytosine (C). RNA is single-stranded and contains uracil (U) instead of thymine.

Base pair: Two complementary nucleotides bonded together at the matching bases (A and T or C and G) to form a double-stranded complex; the length of the DNA is often described in base pairs (bp).

Base pairing: The formation of hydrogen bonds between the nitrogenous bases of two nucleic acid molecules.

Biotechnology: The set of biological techniques developed through basic research and now applied to research and product development. In particular, the use of microorganisms, plant, and animal cells to produce useful materials, such as food, medicine, and other chemicals.

Biotin: A small vitamin used to label nucleic acids for a variety of purposes, including non-isotopic hybridization by chemiluminescence or chromogenic techniques.

Capillary electrophoresis (CE): A technique for separating DNA from a fluid substrate; the sample is injected into a capillary tube, which is then subjected to a high-voltage current that separates its chemical constituents based on charge and size.

cDNA (complementary DNA): DNA synthesized from an RNA template. The single-stranded form of cDNA is an important laboratory tool (e.g., as a probe) for isolating and studying the expression of individual genes.

Chemiluminescence: A non-isotopic hybridization detection technique. Chemiluminescence is the production of visible light by a chemical reaction.

Chromatin: The complex of genomic DNA and protein found in the nucleus of a cell in interphase.

Coding strand: In double-stranded DNA, the strand that has the same sequence as the resulting RNA (except for the substitution of uracil for thymine).

Codon: A triplet of nucleotides in an RNA molecule that specifies the placement of an amino acid during protein synthesis.

Coincidental match: A match between DNA profiles (i.e., from an evidentiary sample and a known reference sample) that occurs by chance.

Complementary DNA (cDNA): DNA enzymatically synthesized *in vitro* from an RNA template, by reverse transcription. cDNA may be single-stranded or double-stranded, as required by the parameters governing a particular assay. The synthesis of cDNA represents a permanent biochemical record of the cellular biochemistry and also provides a means by which that record can be propagated.

Cytoplasm: The cellular contents found between the plasma membrane and the nuclear membrane.

Denaturation (of nucleic acids): Conversion of DNA or RNA from a double-stranded form to a single-stranded form. This can mean dissociation of a double-stranded molecule into its two constituent single strands, or the elimination of intramolecular base pairing.

Deoxyribonucleic acid (DNA): The substance of heredity; a large molecule that carries the genetic instructions that cells need to replicate and to produce proteins. Consists of two long polymers of simple units called nucleotides (A, T, G, and C) with a sugar and phosphate backbone, assembled by a DNA polymerase. *In vivo*, DNA is produced by the process known as replication. DNA can also be synthesized using a variety of *in vitro* methods, such as the polymerase chain reaction.

Deoxyribonucleotide (dNTP): A single unit of deoxyribonucleic acid (DNA) containing a nitrogenous base, a deoxyribose sugar, and a phosphate group.

Dideoxyribonucleotide (ddNTP): A nucleotide that lacks the 3′-hydroxyl (-OH) group on the corresponding deoxyribose sugar. The absence of this hydroxyl group means that, after the ddNTP is added to its complementary nucleotide by DNA polymerase, no additional nucleotides can be added since the phosphodiester bond cannot be created. Used in mtDNA and SNP analyses.

Diploid: Having two complete sets of chromosomes (two of each chromosome). *Compare to* haploid and triploid.

Direct repeats: Identical or closely related sequences present in two or more copies in the same orientation on the same molecule of DNA; they are not necessarily adjacent.

DNA (deoxyribonucleic acid): The substance of heredity; a large linear molecule that consists of deoxyribose sugar, phosphate groups, and the bases adenine, thymine, guanine, and cytosine.

DNA polymerase I: A prokaryotic enzyme capable of synthesizing DNA from a DNA template. The native DNA polymerase I, also known as the holoenzyme or Kornberg enzyme, manifests three distinct activities: 5′ to 3′ polymerase, 5′ to 3′ exonuclease, and 3′ to 5′ exonuclease.

DNA profile: The pattern of band lengths on an autoradiograph (*see* autoradiography) or "peaks" on an electropherogram representing all of the tests to link DNA samples with probes.

DNA sequencing: A technology for determining the order of nucleotides in a specific DNA molecule.

Dot blot analysis: A rapid, quantitative assay for determining the prevalence of a DNA or RNA sequence in a sample. Denatured samples are applied directly to a filter without prior electrophoretic separation. Results are based on signal intensity within the "dot." Dot blot analysis lacks the qualitative component associated with gel electrophoresis. *See* slot blot analysis.

Duplex: The formation of a double-stranded molecule or portion thereof by the base pairing of two complementary polynucleotides.

Dye blob: Following primer synthesis, some "leftover" dye molecules are not removed by post-synthesis purification and are carried through the PCR amplification step and injected onto the capillary to produce "dye blobs" or "dye artifacts" in CE electropherograms (wider than true allele peaks).

Electropherogram: A recording of the separated DNA components of a sample produced by gel or capillary electrophoresis. A photograph or printout of a gel or capillary separation made after electrophoresis, which records the spatial distribution of macromolecules within the gel or capillary.

Electrophoresis: A type of chromatography in which macromolecules (i.e., proteins and nucleic acids) are separated and resolved through a matrix based on their charge.

Ethidium bromide (EtBr): A planar, intercalating agent used to visualize nucleic acids, both DNA and RNA. This dye emits a bright orange fluorescence when UV irradiated; thus, gels that contain samples can be photographed for future reference. Standard ethidium bromide stock solution is 10 mg/ml in water; standard staining concentration is 0.5–1.0 mg/ml.

Ethylenediamine tetraacetic acid (EDTA): A chemical preservative added to biological samples to inhibit the activity of enzymes that are responsible for degrading DNA.

Evidence sample: Also known as questioned sample.

Exclusion: An interpretation or conclusion that eliminates an individual as a potential contributor of DNA to an evidentiary sample. The genotype comparison shows profile differences that can only be explained by the two samples originating from different sources.

Formaldehyde (HCHO): A commonly used denaturant of RNA.

FTA collection card: An absorbent cellulose-based paper that contains chemical substances to inhibit bacterial growth and to protect the DNA from enzymatic degradation. Liquid samples such as blood and saliva are often collected and "spotted" onto the card for short- or long-term storage at room temperature.

Gel electrophoresis: The process of separating DNA by size in an electrical field; the different sized fragments move at different rates through the gel.

Gene: The fundamental physical and functional unit of heredity. A gene is an ordered sequence of nucleotides located in a particular position on a chromosome that ultimately encodes for the synthesis of a polypeptide.

Genetic code: The language in which the instructions of DNA are written. It consists of triplets of nucleotides (codons), with each corresponding to an amino acid or a signal to start or stop protein synthesis.

Genome: The entire chromosomal DNA found in a cell; its size is generally given in the total number of base pairs. In some applications, it may be useful to distinguish nuclear genomic DNA from the mitochondrial genome.

Genomic DNA: Chromosomal DNA.

Genotype: The genetic composition of an individual cell or organism; the total of all the genes present in an individual.

GM9947A: A Epstein-Barr virus-immortalized human lymphoid cell line (of female origin) that is used as a positive control or reference standard in forensic DNA typing.

Haploid: Having one complete set of chromosomes (one of each chromosome, as found in gametes). *Compare to* diploid and triploid.

Haplotype: Refers to the genetic constitution of an individual chromosome. Haplotype may refer to only one locus or to an entire genome. In the case of humans, a genome-wide haplotype comprises one member of the pair of alleles for each.

Heterozygous: Both alleles at a given locus on each of a pair of homologous chromosomes are different, one inherited from each parent. Two distinct peaks or bands are observed at a locus in an electropherogram or by gel electrophoresis, respectively.

Homologous chromosome: Chromosomes that share an identical sequence of genes, but may carry similar or different alleles at the same loci; associate in pairs.

Homozygous: Both alleles are the same (or indistinguishable) at a given locus, one inherited from each parent. One distinct peak or band at a locus is observed in an electropherogram or by gel electrophoresis, respectively.

Human buffy coat: Human blood can be separated into three distinct components by centrifugation. After centrifugation, the first component or top layer contains clear fluid (the plasma), a thin second layer makes up less than 1% of the total volume of blood, and a third layer of red fluid contains most of the anucleated red blood cells. The second or middle layer contains most of the white blood cells and platelets and is called the buffy coat because it is usually buff in hue or white in color. The buffy coat is used to extract DNA from the blood of humans and animals.

Hydrogen bonding: The highly directional attraction of an electropositive hydrogen atom to an electronegative atom such as oxygen or nitrogen. This is the manner of interaction between complementary bases during nucleic acid hybridization. *See also* base pairing.

Image: The document that image analysis software works upon. Image may also refer to the original artwork, graphics, or photograph that is scanned or imported into image analysis software.

Image analysis: An electronic method for the digital capture and storage of an image, accompanied by automated measurement of parameters such as molecular weight, mass, relative abundance, and optical density of various objects in the image (e.g., bands on a gel).

Inclusion: An interpretation or conclusion for which an individual cannot be excluded as a potential contributor of DNA obtained from an evidentiary sample based on the comparison of the known reference sample to the questioned or evidentiary sample.

Inconclusive/uninterpretable: The data do not support a conclusion as to whether the profiles match. This finding might be reported if two analysts remain in disagreement after review and discussion of the data and it is felt that insufficient information exists to support any conclusion.

K562: A continuous human leukemia cell line (of female origin) that is used as a reference standard in forensic DNA typing.

Known reference sample: Biological material (i.e., blood or a buccal swab) for which the identity of the donor (i.e., victim or suspect) is established and used for comparison purposes. Referred to as a K sample or known sample.

Likelihood ratio (LR): The ratio of two probabilities of the same event under different hypotheses where the numerator contains the prosecutor's hypothesis and the denominator contains the defense's hypothesis.

Locus: The precise position or location of a particular gene, and any possible allele, on a chromosome.

Major contributor: The predominant DNA profile from an individual in a sample that has been determined to be consistent with a mixture.

Marker: A very generic term that can refer to any allele of interest in an experiment. Also, marker can refer to a molecular standard.

Masked allele: An allele of a contributor in a mixture that may not be readily distinguishable from the alleles of the other contributor(s).

Minor contributor(s): The lesser DNA profile from an individual in a sample that has been determined to be consistent with a mixture.

Mixture: A biological sample, generally an evidentiary sample, that contains DNA from two or more individuals.

Multiplex PCR: Simultaneous amplification of two or more targets in the same PCR reaction.

Noise: The background signal detected by an instrument collecting data.

No results: No detectable allelic peaks at a given locus(i) above the analytical threshold.

Nucleotide: A subunit or molecule of DNA or RNA consisting of a 5-carbon sugar (ribose or deoxyribose), a nitrogenous base (adenine, cytosine, guanine, thymine, or uracil), and a phosphate group. Nucleotides are the building blocks used to assemble both DNA and RNA.

Oligonucleotide: A short, artificially synthesized, single-stranded DNA molecule that can function as a nucleic acid probe or a molecular primer. *Oligonucleotide* can also refer to a short fragment of RNA.

Palindrome: A segment of duplex DNA in which the base sequences of the two strands exhibit a two-fold rotational symmetry about the central axis. Restriction endonucleases often recognize and cut the DNA at a variety of such palindromic sites.

Partial profile: A DNA profile whereby the typing results obtained are not from all loci tested due to degradation, low template (DNA) concentration, or inhibition of the PCR amplification process.

PCR (polymerase chain reaction): A systematic, primer-mediated enzymatic process for the geometrical amplification of a target DNA sequence. PCR product can be generated from as little as one molecule of target material (DNA or RNA) under optimal conditions.

Peak height ratio: The ratio of two alleles at a given locus that is determined by dividing the peak height of the allele with a lower response (*see* relative fluorescence units or RFUs) by the peak height of an allele with a higher response or RFU value, and then multiplying by 100 to express the peak height ratio as a percentage. Generally used to determine which alleles may be heterozygous pairs in a mixture.

Phenotype: The observable characteristics of an organism or individual.

Photodocumentation: A method for preserving the image of a gel immediately after electrophoresis, or after hybridization with a labeled probe. Media that support photodocumentation include Polaroid film, x-ray film, thermal paper, and digital storage. *See* image analysis.

Polymerase chain reaction (PCR): A primer-mediated enzymatic process for the systematic amplification of minute quantities of specific genomic or cDNA sequences. This technique, which has revolutionized molecular biology, mimics DNA replication. It has the advantage of being a very sensitive technique that can be performed in a short time frame order, amplifying a targeted sequence hundreds of millions to billions of times.

Polymorphism: The quality or character occurring in more than one form.

Primer: A short nucleic acid molecule that, upon base pairing with a complementary sequence, provides a free 3′-OH for any of a variety of primer extension-dependent reactions.

Probability of exclusion (PE): The percentage of the population that can be excluded as potential contributors to a mixture.

Probability of inclusion (PI): The percentage of the population that can be included as potential contributors to a mixture.

Probe: Usually, labeled nucleic acid molecules, either DNA or RNA, used to hybridize to complementary sequences in a library, or which are among the complexity of different target sequences present in a nucleic acid sample, as in the Northern analysis, Southern analysis, or nuclease protection analysis. In forensics, a short segment of DNA is used to detect certain alleles. The probe hybridizes, or matches up, to a specific complementary sequence, allowing for the visualization of the DNA complex, either by a radioactive tag (RFLP) or biochemical tag (HLA DQA1). A single-locus probe marks a specific site (locus), whereas a multilocus probe marks multiple sites.

Pull-up (or bleed-through): When STR amplified products are separated by capillary electrophoresis, the analytic software fails to discriminate between the different dye colors (attached to each allele) used during the generation of the test results. For instance, an allele at a given locus labeled with a blue dye might mistakenly be interpreted as a green signal for a different locus.

Questioned sample: A biological sample recovered or collected from a crime scene or from individuals or objects associated with a crime (referred to as a Q or queried sample).

Random match probability (RMP): The probability of randomly selecting an unrelated individual from a population who could be a potential contributor to an evidentiary sample or profile.

Reference sample: Also referred to as the known reference sample. *See* known reference sample.

Relative Fluorescence Units (RFU): A unit of measurement used in STR analysis that employs fluorescence detection. Samples that contain higher quantities of amplified DNA will have higher corresponding RFU values (e.g., higher peak hights).

Renaturation: The reassociation of denatured, complementary strands of DNA or RNA.

Replication: The formation of an exact copy. DNA replication occurs when each strand acts as a template for a new, complementary strand, formed according to base pairing rules.

Restriction endonuclease: A class of enzymes that recognize a specific base sequence (usually four to six base pairs in length) in a double-stranded DNA molecule and cuts both strands of the DNA at every site where this sequence occurs.

Restriction fragment length polymorphism (RFLP): The presence of variants in the size of DNA fragments produced upon restriction enzyme digestion due to a change in bases. These different sized fragments may result from an inherited variation in the distribution of restriction endonuclease sites. RFLPs are used in the laboratory for human identification or parentage determination.

RFLP analysis: A technique that uses single-locus or multilocus probes to detect variation in a DNA sequence according to differences in the length of segments created by cutting DNA with a restriction enzyme.

Ribonucleic acid (RNA): A chemical found in the nucleus and cytoplasm of cells. A polymer of ribonucleoside monophosphates, synthesized by an RNA polymerase. RNA is the product of transcription and plays an important role in protein synthesis.

Sequence: The order of nucleotides (A, C, G, and T) in a nucleic acid or DNA molecule.

Short tandem repeats: Discrete sequences of DNA that are repeated end on end. These repeating sequences consist of 2–6 base pairs of DNA that are scattered throughout the human genome. Used as molecular markers in forensic DNA analysis, paternity verification, and kinship determination.

Single-nucleotide polymorphism (SNP): A change in the DNA in which a single base or nucleotide differs from the usual base at that position.

Single-source profile: A DNA profile that has been determined to originate from one source or one contributor; based on the number of alleles at all loci tested and the peak height ratio assessments.

Slot blot analysis: A membrane-based technique for the quantitation of specific RNA or DNA sequences in a sample. The sample is usually slot configured onto a filter by vacuum filtration through a manifold (*see also* dot blot analysis). Slot blots lack the qualitative component associated with electrophoretic assays.

Sodium dodecyl sulfate (SDS): An ionic detergent commonly used to disrupt biological membranes and to inhibit RNase.

Stochastic effects: The observation of intralocus peak imbalance or allele dropout that results from disproportionate PCR amplification of alleles in biological samples containing low levels of DNA.

Stutter: During PCR (*see* polymerase chain reaction), the polymerase fails to copy all of the tandem repeats, creating a subpopulation of STR products "shorter" than the true allele. For example, a true allele may contain seven STRs; however, if stutter occurs, a small subpopulation of six STRs would exist at a given locus. In an electropherogram, a stutter will be revealed by the observation of a small peak prior to the true allele peak.

SYBR Green: One member of a new family of dyes for staining nucleic acids. Commonly prepared as a 10,000× stock solution in DMSO, SYBR Green is diluted to a working concentration of 1× in Tris buffer, such as 1× TAE. The advantages of using SYBR Green include reduced background levels of fluorescence, higher sensitivity, and reduced mutagenicity when compared with ethidium bromide. SYBR Green I is used to stain DNA, while SYBR Green II is used to stain RNA.

***Taq* DNA polymerase:** Thermostable DNA polymerase from the organism *Thermus aquaticus*. *Taq* is one of several enzymes that can be used to support the polymerase chain reaction.

Template: A macromolecular informational blueprint for the synthesis of another macromolecule. All polymerization reactions, including replication, transcription, and PCR, require templates; these dictate the precise order of nucleotides in the nascent strand. Primer extension-type reactions cannot proceed in the absence of template material.

Triploid: Having three complete sets of chromosomes (three of each chromosome). *Compare to* diploid and haploid.

Variable number of tandem repeats (VNTR): Multiple copies of virtually identical base pair sequences, arranged in succession at a specific locus on a chromosome. The number of repeats varies from individual to individual, thus providing a basis for individual recognition.

UV light (ultraviolet light): Short-wave, high-energy portion of the electromagnetic spectrum. Because nucleic acids absorb light maximally in the ultraviolet range (260 nm), samples of nucleic acids are stained with dyes (e.g., EtBr) and irradiated with UV light for visualization. *Caution*: UV light is mutagenic and can severely damage the skin and the retina of the eye. Be certain to wear proper eye and skin protection at all times.

Index

T

U

V

W

Y